T0135506

Detection and Generation of Non-Classical Light States from Single Quantum Emitters

Dissertation

zur Erlangung des akademischen Grades
doctor rerum naturalium
(Dr. rer. nat.)
im Fach Physik

eingereicht an der
Mathematisch-Naturwissenschaftlichen Fakultät I
der Humboldt-Universität zu Berlin

von

Thomas Aichele

geboren am 31. Dezember 1974 in Friedrichshafen

Präsident der Humboldt-Universität zu Berlin
Prof. Dr. Jürgen Mlynek

Dekan der Mathematisch-Naturwissenschaftlichen Fakultät I
Prof. Thomas Buckhout, PhD

Diese Arbeit wurde von der Mathematisch-Naturwissenschaftlichen Fakultät I
der Humboldt-Universität zu Berlin 2005 als Dissertation angenommen.

Bibliografische Information Der Deutschen Bibliothek

Die Deutsche Bibliothek verzeichnet diese Publikation in der Deutschen
Nationalbibliografie; detaillierte bibliografische Daten sind im Internet über
http://dnb.ddb.de abrufbar.

ISBN 3-8325-0867-8

Logos Verlag Berlin
Comeniushof, Gubener Str. 47,
10243 Berlin
Tel.: +49 030 42 85 10 90
Fax: +49 030 42 85 10 92
INTERNET: http://www.logos-verlag.de

Zusammenfassung

Die Erweiterung der klassischen Informationsverarbeitung durch quantenmechanische Prinzipien ermöglicht Anwendungen mit neuartigen Eigenschaften. Dazu gehören Quantenkryptographie für sichere Nachrichtenübertragung und Quantencomputer, welche leistungsfähigere Rechenalgorithmen zulassen. Nicht-klassische Lichtzustände, wie zum Beispiel einzelne Photonen und verschränkte Photonenpaare, sind hierbei von besonderem Interesse, insbesondere beim Austausch von Quanteninformationen über große Distanzen. Dabei liefern spontane Zerfälle von angeregten Zuständen, wie sie zum Beispiel in Atomen und Quantenpunkt-Nanostrukturen vorkommen, exakte Ein-Photon-Zustände mit hoher Effizienz.

In dieser Arbeit wurden Quantenpunkte verschiedener Materialsysteme (InP/GaInP und CdSe/ZnSe) auf Emission nicht-klassischen Lichts und dessen optischen Eigenschaften untersucht. Dabei wurde erstmals die Erzeugung einzelner Photonen an einem InP-Quantenpunktsystem, welches für Freistrahl-Experimente besonders interessante optische Eigenschaften besitzt, beobachtet. Die Emission geschieht dabei durch Rekombination von zuvor optisch angeregten Exzitonen. Durch den Einsatz bei tiefen Temperaturen (ca. 4 K) werden Phonon-Wechselwirkungen der Quantenpunkt-Exzitonen minimiert. Die Emission wird mit einem Konfokalmikroskop effizient gesammelt, dessen Auflösung die Detektion einzelner Quantenpunkte ermöglicht.

In Intensitäts-Korrelationsmessungen nach Hanbury Brown und Twiss wurde ausgeprägtes Photon-Antibunching beobachtet, welches die Emission von nur einem Photon pro Zeiteinheit nachweist. Durch Anregung der Quantenpunkt-Zustände mit gepulstem Laserlicht konnten so zeitlich kontrollierte Ein-Photon-Zustände erzeugt werden.

Mithilfe von Fourier-Spektroskopie wurde die Kohärenzlänge der Photonen in Abhängigkeit von äußeren Parametern, wie Temperatur und Anregungsleistung, aufgenommen. Vergleiche mit Lebensdauermessungen lieferten Aussagen über die Reinheit der spektralen Photon-Mode und über Dekohärenzprozesse in den Quantenpunkten.

Zerfallskaskaden mehrerer Exzitonen in Quantenpunkten führen zur Emission von Photonenpaaren und -triplets unterschiedlicher Wellenlänge. Unter geeigneten Bedingungen wird sogar die Erzeugung von polarisationsverschränkten Photonenpaaren vorausgesagt. Im Rahmen dieser Arbeit konnten durch Kreuzkorrelationsmessungen zwischen den Detektionsereignissen der beiden Photonen eines Paares deutliche Korrelationen der individuellen Emissionszeiten gemessen werden. Dadurch konnten Informationen über die Natur der Quantenpunktzustände und über die Dynamik der Übergänge gewonnen werden. Polarisationsabhängige Messungen gaben erste Hinweise auf Korrelationen des Polarisationszustands der beiden Photonen.

Die Photonenpaare wurden in einem weiteren Experiment durch Ausnutzung ihrer unterschiedlichen Wellenlängen in einem Interferometer effizient aufgetrennt. Dadurch konnten die beiden Photonen unabhängig genutzt werden und, ähnlich dem Multiplexing in der klassischen Kommunikationstechnik, eine Erhöhung der Photonenrate erreicht werden. Es gelang hierbei zum ersten Mal, Multiplexing im Intensitätsbereich einzelner Photonen in einem Quantenkryptographie-Experiment zu demonstrieren.

In einem weiteren Projekt wurden mikrometergroße Spitzen auf einem Halbleitersubstrat untersucht, welche nur einzelne oder wenige Quantenpunkte auf ihrer Oberfläche haben. Diese Spitzen eignen sich sowohl als aktive Sonden in der Mikroskopie als auch für die gezielte Positionierung von Quantenpunkten an Mikroresonatoren. Erste Messungen zeigten Einzelphoton-Emission von Spitzen mit nur einem Quantenpunkt und gaben Hinweise auf Kopplung der Emission an Moden eines Mikrokugel-Resonators.

Abstract

The extension of classical information processing by quantum-mechanical principles allows applications with novel properties. Among them are quantum cryptography, for secure information exchange, and quantum computing, which makes improved algorithms possible. For this, non-classical light states, like single photons and entangled photon pairs, are of special interest, for instance for transmission of quantum information over large distances. The spontaneous decay of excited quantum states, as it is happening in atoms and quantum dots, provides single-photon states with high efficiency.

In this thesis, quantum dots of different material systems (InP/GaInP and CdSe/ZnSe) were investigated with respect to emission of non-classical light and their optical properties. For the first time, the generation of single photons with the InP quantum dot system, which is especially interesting for free-beam experiments, was observed. In quantum dots, photon emission originates from the decay of optically excited excitons. Operation at cryogenic temperatures (around 4 K) minimizes phonon interactions. The emission is efficiently collected using a confocal microsope with a resolution sufficient for detection of single quantum dots.

The observation of pronounced photon anti-bunching in intensity correlation measurements following Hanbury Brown and Twiss proves the emission of only one photon per time unit. Pulsed excitation of the quantum dots generates single-photon states on demand with a well-controlled emission time.

By using Fourier spectroscopy, the coherence length of the photons in dependence of external parameters, like temperature and excitation intensity, was measured. A comparison with lifetime measurements gives information about the photon's mode purity and about decoherence processes in quantum dots.

Decay cascades of multiple quantum dot excitons lead to the emission of photon pairs and triplets of different energy. Under certain conditions, even the generation of entangled photon pairs is predicted. In this work, cross-correlation measurements between detection events of photon pairs were performed, showing pronounced correlations of the different emission times. These measurements give information about the nature of the quantum dot states and about the dynamics of the decay cascades. Polarization dependent measurements gave first evidence of correlations between the polarization state of the two photons.

In another experiment, the photon pairs were efficiently separated in an interferometer, making use of their different wavelengths. In this way, the two transitions in the cascade can be used as two independent single-photon sources and an increase of the photon generation rate could be achieved, similar to multiplexing in classical communication. For the first time, multiplexing on a single-photon level could be demonstrated in a quantum cryptography experiment.

In a further project, micrometre sized tips on a semiconductor substrate with only a few quantum dots on their surface were characterized. Such tips are useful as active probes in microscopy and for controlled positioning of single quantum dots to microcavities. First measurements revealed single-photon emission from tips with only one quantum dot and gave evidence for coupling of the emitted light to the modes of a microsphere cavity.

Contents

Contents

1. Introduction

Photons are the most fundamental excitations of the quantized electro-magnetic field. Thus, they are important research objects in quantum mechanics. Their investigation helped to get a deeper understanding of the phenomenon *light*, like, for instance the explanation of the spectrum of black body radiation, and led to experiments, in which the fundament of quantum mechanics was tested [AGR82].

At the beginning of the eighties, first proposals were made, in which conventional, classical applications in the field of information processing were extended into the quantum world: In 1982, R. P. Feynman suggested to use quantum mechanical states to replace classical bits in a computer [Fey82]. Originally, his idea was to simulate the time evolution of quantum mechanical systems itself. But soon algorithms were discovered, that proved the superiority of such quantum computers to classical computers in certain situations, such as the factorization into prime numbers [Sho97] or the search of elements in an unsorted set [Gro97]. Both applications make use of the fact, that a quantum bit (or qubit) does not only exhibit discrete values, say 0 or 1, but can also form a continuous superposition $\alpha|0\rangle + \beta|1\rangle$ or, for several qubits (a quantum register), even occupy an entangled state. Handling such quantum registers allows the simultaneous exploration of different values in the register. This quantum parallelism enables the computation of a large data space at a given number of operations.

The practical implementation of such a computer however faces enormous and yet unsolved experimental and theoretical challenges. The main difficulty is to maintain quantum coherence during the whole computation process. This means, that the qubits have to be coupled among each other, as well as they have to be completely decoupled from the outside word. Further demands are the possibility to scale up the quantum computer to many qubit systems and the ability to properly initialize and read out the qubits [DiV00]. It was shown that, for a universal operation (that is, to carry out all possible algorithms), a set of one-qubit operations and only one type of two-qubit gate, the controlled-NOT (C-NOT) gate, has to be implemented. Especially the two-qubit gate turned out to be difficult to realize, since it requires the controlled switching on and off of coherent interaction of two quantum systems.

Until today, various approaches for qubit implementations were followed. Basic quantum algorithms were successfully implemented using nuclear spin states [CVZ+98] and motional and electronic states of single ions [GRL+03]. Other promising attempts are spin states of quantum dot excitons [LWS+03], charge population states in coupled quantum dots [BHH+01], qubits based on charges in superconducting islands [YPA+03] and energy levels in Josephson junctions [YHC+02]. In 2001, Knill et al. [KLM01] proposed the implementation of a purely photonic C-NOT gate using solely linear optics with single-photon states as qubits. Single photons are appealing because of the ease with which interference can be observed. Single photons are also interesting as transmitters of quantum information between stationary, matter-based qubits.

A second important application of single photons appeared in 1984, when Bennett and Brassard [BB84] discovered, that data can be transmitted without the possibility of eavesdropping, if encoding information onto a quantum state of single particles. For transmission over larger distances, the photon is currently the only reasonable quantum information carrier. The use of entangled photon pairs leads to improved protocols [Eke91, BGT+99] and to proposals for quantum repeaters [BDC+98], that may allow the purification of quantum state entanglement at regular distances along a communication channel. In this way, decoherence does not limit the communication distance any more.

Single photon sources

In spite of their fundamental character, single photons are not easy to generate. As photons obey the Bose-Einstein statistics, most natural light sources tend to emit multiple photons instead of individual particles per time unit. Figure 1.1 shows probability distributions for measuring a certain amount of photons in a time interval for different example light fields. Thermal light fields (a), such as sun light or light from a bulb, have a Bose-Einstein photon number distribution. In such a field, photons appear bunched together and higher photon numbers become possible. Even laser light, (b), which possesses the narrowest classically obtainable photon statistics, shows a Poissonian distribution $p_n = \exp(-\mu)\mu^n/n!$ with average photon number μ. But for the quantum applications mentioned before, light fields with more than one photon reduce the quality of operation: For photonic quantum gates, but also quantum repeaters and quantum teleportation, multi-photon states may lead to wrong detection events that cause wrong interpretations of the outcome of a quantum operation. In quantum cryptography an eavesdropper may split off additional photons to gain at least partial access to the transmitted key (see also chapter 7).

In contrast to this, an ideal single-photon source has a vanishing probability to measure other than exactly one photon at a time ($p_n = \delta_{1,n}$). Such sub-Poissonian distributions (with a width narrower than that of a Poissonian of the same average photon number) are known to be non-classical and have to be described by means of quantum mechanics. Realistic single-photon sources undergo various loss mechanisms, like emission into uncontrolled optical modes or absorption, so that a more realistic photon number distribution has a certain zero-photon probability, like the one in figure 1.1(c).

For a photon source it is not sufficient to just suppress all higher photon numbers. For useful operation, a high efficiency of photon generation and the ability to trigger the emission time is also of importance. Additionally to the single-particle character,

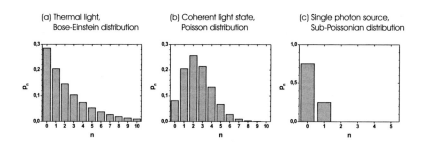

Figure 1.1.: *Photon number distributions of (a) thermal light, (b) a coherent state and (c) a single-photon source (with 0.25% efficiency).*

in many experiments the purity of the spatial and temporal modes is required, as well. This is especially the case in situations where two-photon interference following Hong et al. [HOM87] is observed, as in linear optical quantum gates [KLM01].

The easiest way to approximate single-photon states is to use highly attenuated laser pulses: Due to their Poissonian photon number distribution, the multi-photon probability scales linear with the mean photon number, $p_{\geq 2} \approx \mu p_1/2$, which approximates a single-photon state for $\mu \ll 1$. However, the single-photon efficiency scales in the same way, $p_1 \approx \mu$ for $\mu \ll 1$, which makes this method highly inefficient.

Today, a widely used method to generate single-photon states is spontaneous parametric down conversion [MW95]. In this technique, a non-linear crystal is pumped by a strong laser field, whereupon, with a certain probability, a laser photon is down-converted to a photon-pair of half frequency. The detection of the first photon post-selects the second efficiently in a well defined mode. This predestinates this method for experiments which need a high *gated* photon efficiency and mode purity [LHA+01]. With alterations of this technique, other non-classical states can be generated, like entangled photon-pairs [KMW+95] or photon-added coherent states [ZVB04]. However, due to the stochastic nature of the down-conversion process, the high photon efficiency can only be maintained by gating, whereas only a limited overall efficiency is offered. On the other hand, an increase of pump power to improve the photon rate leads to an increased probability to generate two photon pairs. These facts limit the potential of this method for future commercial quantum applications, where high single-photon rates are needed.

Single quantum emitters

A quantum emitter is generally defined as a quantum system that is capable of radiative optical transitions. When observing the spontaneous decay of a single excited quantum

11

emitter, the emission of a single photon is expected. When suppressing non-radiative decay mechanisms, they can act as principally 100% efficient single-photon sources. To attain this high efficiency, a strong control of the spatial emission mode is required, which sets a technical but no fundamental limit to today's maximally achievable photon efficiency. The variety of systems offered by nature allows a multitude of possible experiments and implementations of photonic applications.

Discrete electronic transitions in atoms are the most self-evident systems and were the first to be investigated in 1977 by Kimble et al. [KDM77]. In recent experiments, single atoms [KHR02, MBB+04] and single ions [KLH+04] were coupled to microcavities to control the emission time and mode of the photon. The emission of an isolated single atom is characterized by a high mode pureness, which is important for any experiments based on two-photon interference. Atom decay cascades can also be used for entangled-photon generation [AGR82]. However, the complexity of today's atom traps restricts this system to fundamental experiments.

Transitions in single molecules and single nanocrystals have also turned out to emit single photons [BLT+99, LM00]. Nanocrystals are semiconductor crystals with sizes of a few nanometres, which are chemically produced as colloids. Similar to quantum dots (see below), nanocrystals show discrete energy levels, in contrast to bulk crystals, leading to single-photon transitions [MIM+00]. Molecules and nanocrystals show similar properties with respect to single-photon emission. Both systems can be operated even at room temperature, making them cheap and easy to handle. Their drawback is their susceptibility for photo-bleaching due to chemical destruction and for blinking. The latter describes the effect of interrupted emission even on large time scales due to the presence of long-lived dark states.

For experiments, where the single-photon statistics is the only important task, diamond defect centres are advantageous. These are impurities in a diamond crystal, where two neighbouring carbon atoms are replaced by a nitrogen atom and a vacancy. These structures show room-temperature single-photon emission without optical instabilities like blinking and bleaching [KMZ+00, BKB+02]. The disadvantage of this system is the broad optical spectrum of the room-temperature fluorescence together with comparably long lifetimes (12 ns).

This work concentrates on single-photon generation using self-assembled single quantum dots. Quantum dots are few-nanometre sized semiconductor structures showing discrete electronic energy levels, in contrast to energy bands in bulk semiconductors.[1] To suppress phonon interaction and thermal ionization, quantum dots mostly need to be operated at cryogenic temperatures, but experiments at increasingly higher temperature have also been reported [SMP+02, Mir04]. The advantage of quantum dots is,

---

[1]Although colloidal nanocrystals are also quantum dots by this definition, to avoid confusion, in this work, the terminology of *quantum dots* is restricted to quantum dot structures on semiconductor substrates.

that they combine narrow spectral lines and short transition lifetimes. In contrast to the similar nanocrystal systems, quantum dots are optically very stable, as they usually show no blinking and have a long working lifetime. Quantum dots also offer the possibility of electric excitation [YKS+02] and of implementation in integrated photonic structures. Due to the variety of possible materials, quantum dots have shown single-photon emission throughout the visible and infrared spectrum (see references [SMP+02, BPK+04, SPS+01, SPS+01, MRG+01] and the experiments in this thesis). Moreover, it was proposed to use quantum dot multi-photon cascades for the generation of entangled photon-pairs [BSP+00].

Outline of this thesis

The topic of this thesis is the generation, detection and characterization of single-photon light states emitted by single quantum dots. These light states are in particular single-photon states but also photon pairs and triplets from cascade decays. The suitability of the generated light fields for quantum optical applications was tested in different characterization and demonstration experiments. This thesis is structured in the following way:

Chapter 2 gives background information over two main topics, this work is engaged in: First, the verification of single photon emission by measuring the second order coherence function, using the Hanbury Brown-Twiss configuration, is described. Second, the relevant physical properties of quantum dot systems and their production process are explained.

The experimental setup which was used in the described experiments is presented and characterized in **chapter 3**. Variations and add-ons in specific experiments will be explained in the according section.

The first step for all succeeding experiments is the verification of single-photon emission. In **chapter 4**, both pulsed and continuous single-photon emission from different quantum dot systems are presented. A model describing the photon emission of the quantum dots is discussed and evaluated using Monte Carlo simulations.

The spectral and coherence properties of single photons are vital for the outcome of many quantum optic experiments. The results of Fourier spectroscopic measurements to acquire these characteristics are shown in **chapter 5**. These measurements describe the wave nature of light, in contrast to the particle-like characteristics revealed in the previous chapter. This circumstance is argued by arranging an experiment to simultaneously detect interference and anti-bunching.

Chapter 6 is dedicated to the investigation of multi-photon cascades. Strong time-correlations between successively emitted photons over a three-photon cascade have been observed. The measurements were complemented by an analytic solution of a rate model. First results exposing polarization dependent correlations are shown.

In **chapter 7** the applicability of the single-photon source is demonstrated. An experiment was conceived and realized to enable multiplexing, similar to that in classical communication, on the single photon level. Quantum key distribution is accomplished to demonstrate the usability of this technique.

Finally, **chapter 8** gives an outlook to further experiments based on the presented work. This also includes first results of a project using single quantum dots grown on pre-structured micro-tips for coupling of dots to microcavities.

2. Physical Foundations of Single-Photon Generation with Single Quantum Dots

2.1. Verification of single-photon emission

As already mentioned in the introduction, different light sources show different photon statistics even when operated at the same intensity. These characteristics play an important role for many photonic applications. In this section, the background for the verification of single-photon states is discussed.

The acquisition of intensity correlations has turned out to be a standard method for testing single-photon emission: The intensity correlation of a light field is detected at two points in time, resulting in the second order coherence function[1] (or simply (auto-)correlation function), which was first introduced by Glauber [Gla63]. In quantum mechanical description, it has the form:

$$g^{(2)}(t, t') = \frac{\left\langle : \hat{I}(t)\hat{I}(t') : \right\rangle}{\left\langle \hat{I}(t) \right\rangle^2}, \tag{2.1}$$

where $::$ denotes normal ordering of the operators. This function is proportional to the joint probability of detecting one photon at time t and another at t'. Usually, one is concerned with stationary fields, in which case the ensemble averages are time independent and the correlation function depends only on the time difference $\tau = t - t'$:

$$g^{(2)}(\tau) = \frac{\left\langle : \hat{I}(0)\hat{I}(\tau) : \right\rangle}{\left\langle \hat{I}(0) \right\rangle^2}. \tag{2.2}$$

This function has several characteristic properties: As each random process is assumed to become uncorrelated after a sufficiently long time scale, the correlation function tends to a normalization value of 1 for long times. It can be further shown [MW95] that for

---

[1]The first order coherence function is obtained by detecting correlations of the field amplitudes, when measuring interference.

all classical fields $g^{(2)}(0) \geq 1$ and $g^{(2)}(0) \geq g^{(2)}(\tau)$. From this, it immediately follows that for classical light fields, no function values smaller than 1 can be observed.

The case $g^{(2)}(0) > 1$ is especially characteristic for thermal light sources. In this case, the photons in the field appear bunched, which means that the probability to detect a second photon soon after a first one is increased (figure 2.1(a)). As the width of the bunching peak is approximately the inverse of the spectral bandwidth, bunching of thermal sources can only be observed when the linewidth is narrow compared to the detection time resolution. Bunching at larger time scales can also be observed with single-photon sources, which indicates correlations between successively emitted photons (like for example in Rabi oscillations [KDM77]).

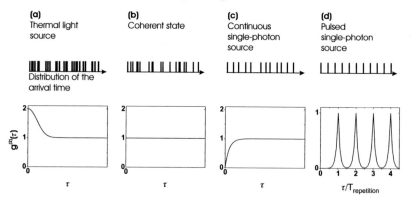

Figure 2.1.: *Top: illustrative distribution of the photon arrival time, bottom: second order coherence function $g^{(2)}(\tau)$ of (a) a thermal light source (e.g. a light bulb), (b) coherent light (laser light), (c) a continuously driven single-photon source and (d) pulsed single-photon source.*

For coherent light fields, such as continuous laser light, $g^{(2)}(\tau) = 1$ for all τ, which indicates a Poissonian photon number distribution. In fact, in this special case, the field is referred to as coherent in second order. For such a Poissonian process, the photons arrive independently (figure 2.1(b)).

When, however, the probability to detect a second photon soon after a first detection is reduced compared to an independent process, then $g^{(2)}(0) < 1$ (figure 2.1(c)). This effect is called anti-bunching. As mentioned before, this case is reserved to non-classical states with sub-Poissonian photon statistics. For photon number states $|n\rangle$, with exactly n photons, $g^{(2)}(0) = 1 - 1/n$ and in the special case of a single-photon state $(n = 1)$, $g^{(2)}(0) = 0$. Thus, when the quantum emitters which generate this field emit independently, measuring $g^{(2)}(0)$ allows counting of a smaller amount of emitters. For statistical mixtures of one- and two-photon states (or more), intermediate values can also be obtained.

In the case of a pulsed source, the second order coherence function possesses a peaked structure. Here, a missing peak at $\tau = 0$ indicates the generation of only one photon per pulse (figure 2.1).

2.1.1. Measurement of the second-order coherence function

The straightforward method to measure the second order coherence function would be to simply note the times of a photodetector's detection events and to compute the correlation function according to equation (2.2). However, this approach prevents the measurement of time scales smaller than the detector's dead time. For example, avalanche photodetector modules, which offer the highest detection efficiencies in the visible spectrum, have dead times around 50 ns [Las]. To overcome this problem, the arrangement in figure 2.2, using two photodetectors monitoring the two outputs of a 50:50 beam splitter, is chosen. Such a setup was originally used by Hanbury Brown and Twiss to detect the *spatial* correlation function, with which they determined the diameter of stars [HT56]. With the Hanbury Brown-Twiss arrangement, the second detector can be armed right after the detection event of the first. For comparably small count rates, the case, where the first detector is already armed while the second one is still dead, can be neglected. Losses, like photons leaving the wrong beam splitter output or undetected photons, simply lead to a global decrease of the measured, un-normalized correlation function.

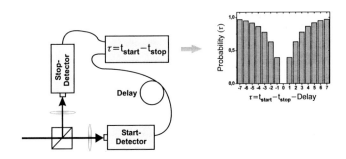

Figure 2.2.: *Scheme of the Hanbury Brown-Twiss setup.*

Technically, it is very difficult to acquire absolute detection times with a resolution in the nanosecond regime. Additionally, the sheer amount of detection events needed to reach a reasonable statistics ($\approx 10^7$ events for a count rate of 10^5 s^{-1}) makes the computation of the correlation function very time-consuming. Instead, only the time differences between the detection events are usually registered and binned together in a histogram. An electronic delay shifts the time origin and enables the observation of asymmetric cross-correlation functions (see chapter 6). Using a time-to-amplitude converter, time differences can be measured very precisely. This method has the additional advantage, that the evolution of the measurement can be tracked online. But one has to keep in mind, that the such observed function $d(t)$ differs from the original second order coherence function $g^{(2)}(t)$. Now, the probability density to measure a time difference at time t is given by:

$$
\begin{aligned}
d(t) \;=\; & \text{(Prob. density to measure a stop event at time } t \text{ after a start event at time 0)} \\
& \times \text{(Prob. that no stop detection has occurred before)} \\
=\; & (T g^{(2)}(t) + r_D)\left(1 - \int_0^t d(t')\, dt'\right),
\end{aligned}
\tag{2.3}
$$

where the total setup transmission T was introduced to account for possible photon losses and r_D describes the detector dark count rate. Only if the average arrival time of the photons $\bar{t} = r_c^{-1}$, with the photon count rate r_c, is much smaller than the observed time delay t between a start and a stop event, the probability, that no stop detection has occurred before, is approximately 1 and $g^{(2)}(t)$ approaches $d(t)$.

An exact solution of the above integral equation can be achieved by substituting $D(t) = \int_0^\infty d(t')\, dt'$, which gives:

$$
\frac{d}{dt} D(t) = (T g^{(2)}(t) + r_D)(1 - D(t)),
\tag{2.4}
$$

which is solved by

$$
D(t) = 1 - C \exp\left(-\int_0^t (T g^{(2)}(t') + r_d)\, dt'\right).
\tag{2.5}
$$

The initial condition $D(0) = 0$ (no stop event occurred before $t = 0$) specifies the integration constant $C = 1$. Note, that $\lim_{t\to\infty} D(t) = 1$, which indicates that the stop event is detected at least at some time. Thus, the detected function can be deduced:

$$
\begin{aligned}
d(t) \;=\; & (T g^{(2)}(t) + r_D) \exp\left(-\int_0^t (T g^{(2)}(t') + r_d)\, dt'\right) \tag{2.6} \\
\approx\; & \text{const.} \times e^{-(r_c + r_D)t} \qquad \text{for large } t. \tag{2.7}
\end{aligned}
$$

The second line indicates, that for long time differences, the measured histogram decays exponentially on a time scale given by the detector count rate. Figure 2.3 shows such a large-time measurement of a single photon source (see also the inset).

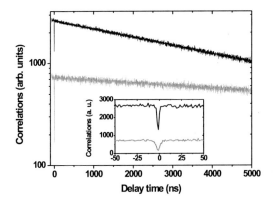

Figure 2.3.: *Correlation measurement of a spectral line of a single quantum dot over a time scale large, compared to the average time between detection events (7.7 µs for the black and 20 µs for the gray curve). The logarithmic scale emphasizes the exponential behaviour. The dip at time 0 (see zoom in the inset) indicates single-photon emission.*

2.2. Self-assembled quantum dots

It is well-known that electrons, which are confined in a potential trap with extensions similar or small compared to that of the electron wave-function, exhibit a discrete energy spectrum.

Quantum dots are intermediate systems on the evolution from a single atom to a solid. They are semiconductor structures with very small spatial dimensions, surrounded by higher band-gap material. In such a system, electrons in the conduction band and holes in the valence band are strongly confined in all three spatial dimensions (see figure 2.4). This leads to a discretization of the energy level scheme, which makes quantum dots in many ways similar to atoms rather than to bulk semiconductors.

2.2.1. Electrical and optical properties of quantum dots

For the calculation of electronic states in quantum dots, several schemes have been used with different levels of sophistication [BGL88]. As an example, here one of the simplest models, an electron in a cubic potential with infinite barriers, is described. In the

19

effective mass approximation, the wave function of a single electron in the conduction band, which experiences a slowly varying potential, is given by:

$$\psi_e(\mathbf{r}) = \phi(\mathbf{r})u(\mathbf{r}), \tag{2.8}$$

where the Bloch function $u(\mathbf{r})$ is periodic with the lattice periodicity. The slowly varying envelope function $\phi(\mathbf{r})$ obeys the Schrödinger equation:

$$\left(-\frac{\hbar^2}{2m^*}\nabla^2 + V(\mathbf{r})\right)\phi(\mathbf{r}) = E\phi(\mathbf{r}), \tag{2.9}$$

with the effective electron mass m^*. For a cubic quantum dot of size L with infinite barriers the effective potential has the form:

$$V(\mathbf{r}) = \begin{cases} -V_0, & |x|, |y|, |z| \leq L/2 \\ \infty, & \text{else} \end{cases}. \tag{2.10}$$

When solving this, the energy eigen-values of this systems are found to be:

$$E_{lmn} = \frac{\pi^2\hbar^2}{2m^*L^2}(l^2 + m^2 + n^2), \quad \text{for } l, m, n \in \{1, 2, 3, \ldots\}. \tag{2.11}$$

It can be seen that only discrete energies are allowed, comparable to the situation in atoms. Higher dimensional structures as quantum wires (two-dimensional confinement) and quantum wells (one-dimensional confinement) also have quantized $\mathbf{k}$-vectors along the confinement direction. For finite barriers and small sizes, only the first few states are considered, while above the potential barrier a continuum of energy levels is present. Although, here, many simplifications were made compared to more realistic dot geometries, this model is suitable to give a qualitative understanding and to demonstrate the discrete energy scheme. For more realistic potentials (see the book of Bimberg et al. [BGL88]), the degeneracy of the levels might be changed or lifted.

When the quantum dot is occupied with several quasi-free charge carriers (electrons or holes), Coulomb interaction has to be taken into account. While equally charged carriers suffer a repulsion, for an electron–hole pair the energy of the system is lowered and an *exciton* is formed (figure 2.4(a)). The recombination of the exciton leads to the emission of a single photon. One distinguishes two typical regimes. If the extension of the quantum dot clearly exceeds the exciton Bohr radius, the centre-of-mass motion is quantized by the confinement potential, while the relative carrier motion is dominated by the Coulomb interaction. This case is called the weak-confinement regime. In the strong-confinement regime the dot radius is smaller than the exciton Bohr radius, and the kinetic energy, due to size quantization, is the dominant energy contribution.

In the same way, two electron–hole pairs form a biexciton, but with in general different energy due to Coulomb interaction (figure 2.4(b)). When decaying, first one electron–hole pair recombines, which leads to the emission of one photon and to a remaining

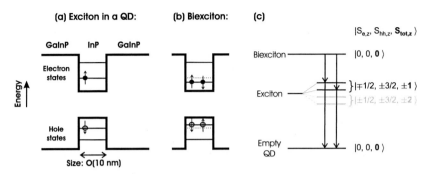

Figure 2.4.: *Excitations in a quantum dot: (a) Exciton formed by an electron–hole pair, (b) The biexciton containing two electron–hole pairs with generally a different energy than the exciton. (c) Schematic term scheme for the exciton and biexciton decay cascade. The two dark excitons are indicated by gray lines.*

exciton in the quantum dot, which can lead to a second emission of a photon with different wavelength (figure 2.4(c)).

The excitation of a quantum dot can be performed either electrically (see for example references [YKS+02, XWC04]) or optically, as it is done in this thesis. If the ground state of the exciton is pumped resonantly, absorbed and emitted light have the same wavelength, making it impossible to separate luminescence and stray light in continuous wave (cw) experiments. For excitation of the exciton close to resonance, an electron–hole pair is created in the higher settled states directly inside the quantum dot. Such an excited exciton state relaxes very quickly and non-radiatively to its ground state (for InP quantum dots this happens on the order of some 10 ps [VMR+96]). In another resonant technique, two-photon excitation of the biexciton is performed [FBA+04]. When exciting at energies higher than the band gap of the barrier material, free charges are created directly in the conduction and valence band of the bulk, respectively, and finally captured by the quantum dot, where they relax quickly to the exciton ground state. Figure 2.5 shows a microscope image of the photoluminescence of a set of quantum dots. The image was taken through a bandpass filter to suppress excitation stray light.

Due to the Pauli exclusion principle, each electron state can be occupied by at most two electrons corresponding to the two spin states $S_{e,z} \pm 1/2$. In low dimensional systems the dispersion relation is changed with respect to a bulk crystal, so that heavy holes (with $S_{hh,z} = \pm 3/2$) possess a lower energy than light holes ($S_{lh,z} = \pm 1/2$). Thus, in the lowest state, the quantum dot can also be occupied by only two holes leading to four combinations of the electron and hole spin in the exciton ground state with total spin $S_{tot,z} \in \{-2, -1, 1, +2\}$. In the biexciton ground state both spin states of the electrons

21

Figure 2.5.: *Micro-photoluminescence image of InP quantum dots in GaInP.*

and holes are occupied, resulting in a single state with $S_{\text{tot},z} = 0$. As for radiative transitions, the change of spin has to be carried away by the photon, two of the exciton states (with total spin ± 2) are dark states and participate neither in the biexciton nor the exciton decay, as symbolized in figure 2.4(c). A more extended overview of the level scheme of quantum dot excitons can be found in [DGE+00].

2.2.2. Fabrication of self-assembled quantum dots

There are various techniques to achieve zero-dimensional semiconductor structures. One of the first realizations of quantum dots were nanocrystal inclusions in glass [EOT80] or in colloidal solutions (see for example [MNB93, AWJ+00]). These dots are usually made of II-VI compounds (for instance CdSe). They emit at room temperature and are available for the whole visible spectrum.

There is a variety of production methods, in which quantum dots were elaborated by starting from higher dimensional semiconductor hetero-structures [BGL88], like etching pillars in quantum well systems or forming intersections of quantum wells or quantum wires. Also the growth of nano-structures on patterned substrates, such as grooves and pyramids led to successful quantum dot formation [JPSG+] and single photon emission [BPK+04]. These fabrication methods allow a high degree of control over the positioning which is advantageous if one wants to couple the quantum dot to microcavities and photonic devices.

So-called natural quantum dots are formed by width fluctuations mainly of quantum wells but also in nanotube systems. In this environment the excitons are trapped in broader regions of the quantum well, where the confinement energy is lowered, so that a potential minimum is formed. Such excitons exhibit large oscillator strengths leading to short radiative lifetimes [BEG+98, RSL+04], as the lateral size of natural quantum dots is usually much larger than the exciton Bohr radius .

The experiments described in this thesis are performed on self-assembled quantum dots. These quantum dots are fabricated by epitaxial growth of one type of crystal on top of another. If the lattice constants of the two materials are very close (such as AlAs and GaAs) large thicknesses of the two layers can be achieved, which is called Frank-van der Merwe growth. However, if the lattice constants differ noticeably, dislocations due to strain will be created and material islands are formed to minimize the strain. A thin layer, which is known as the wetting layer, will remain, completely covering the substrate. This growth mode is called Stranski-Krastanov growth. The wetting layer forms a quantum well, which usually shows photoluminescence below the quantum dot emission wavelength. Typically, quantum dots are additionally overgrown by the substrate material. This gives a protection to the nano-structures but also influences its optical properties by modifying the barrier bandgap and by introducing further strain [JHH+03].

To produce such structures, crystals are grown atomic layer by atomic layer using epitaxy. There are different epitaxial techniques. In Metal-Organic Vapour Phase Epitaxy (MOVPE), molecules containing the atoms to be deposited on a substrate are brought to the surface in a carrier gas (Hydrogen or Nitrogen), where they decompose due to heat and collisions with the carrier gas, leaving the atoms on the substrate surface. The remaining molecules detach from the surface and are flushed out by the carrier gas. In Molecular Beam Epitaxy (MBE), atoms or molecules are delivered to the substrate through an ultra-pure, ultra-high vacuum atmosphere. The heated substrate surface allows the arriving particles to distribute themselves uniformly across the surface to adopt the crystallographic arrangement of the substrate. MBE can produce very clean samples as the growth is done under ultra-high vacuum conditions. In Chemical Beam Epitaxy (CBE) the sample is also placed in a very low pressure chamber and a beam of reactants is directed to the substrate, where the gas cracks, leaving the desired atoms on the surface.

Figure 2.6 displays a schematic representation of the wavelength emission ranges for several Stranski-Krastanow quantum dot material systems. Up to now, single-photon generation has been achieved with InAs/GaAs, InGaAs/GaAs, InP/GaInP and CdSe/ZnSe quantum-dot systems.

The InAs/GaAs system is by far the most studied of all quantum-dot systems. Most work on single photon emission reported to date has been done on InAs dots emitting in the 900-950 nm range [SPS+01, SPS+01, YKS+02], but also at 1250 nm [MRG+01] and 1300 nm [ZZB+04]. Photon correlation measurements at these wavelengths require good infrared single-photon detectors. Single-photon generation on demand at 1300 nm will be very useful for quantum cryptography with optical fibres. InAs dots imbedded in InP have been shown to be efficient emitters in the infrared, but no single dot optical studies have been reported, so far. This material system may be used to generate single

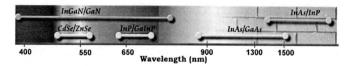

Figure 2.6.: *Schematic representation of the wavelength ranges accessible with different Stranski-Krastanow quantum dot material systems. The region inaccessible to silicon detectors is indicated by a brick wall.*

photons at 1.55 μm, where optical fibres have their lowest absorption. InGaN quantum dots imbedded in GaN have the potential to cover the complete visible range.

InP dots in GaInP have been used in this work to generate single photons in the 640-690 nm range as well as photon pairs and triplets. In principle, this material system could be used to generate single photons between 620 nm and 750 nm. The emission wavelength of these dots fits to the maximum efficiency of silicon avalanche photodiodes, where a detection efficiency in excess of 70% can be reached at around 700 nm, which makes them especially interesting for free space experiments.

Another material system used in this work are CdSe quantum dots in ZnSe. They have been used to generate single green photons at around 500 nm (see also [SMP+02]). One great advantage of such II-VI typed quantum dots over the previously described III-V systems is their short lifetime, which reduces the probability of decoherence during the emission process and enables the generation of single photons on demand with a small time uncertainty, suggesting that the maximum single-photon-emission rate can be much higher than for III-V dots. This system has also a larger energy splitting between the exciton and the biexciton than for the InAs/GaAs material system, which is useful to achieve a better filtering of the exciton emission, making operation at higher temperatures possible. The refractive index of ZnSe is lower than for GaAs, which reduces photon losses due to total internal reflection at the sample surface.

3. Experimental setup

In order to perform experiments with single quantum dots, several requirements concerning the setup have to be fulfilled: As self-assembled quantum dot samples show high quantum dot densities of $10^8 \ldots 10^{11}$ cm^{-2}, a high optical resolution is desired to spatially select one quantum dot or at least only as few as possible. At the same time, a high collection is preferred to loose a minimum amount of photons. These two requirements can be combined by choosing a micro-photoluminescene microscope setup. With this well-established technique, the photoluminescence is studied on the nanometre scale [BKB+02, JVP+03, ZBJ+01, SPS+01].

Figure 3.1 shows a scheme of the experimental setup. The sample is mounted inside a continuous-flow liquid Helium cryostat. Optical access for excitation of the sample and collection of the emitted light is provided through a thin glass window. The sample is excited by either a pulsed or a continuous wave (cw) laser. Light is sent into the microscope objective via a dichroic mirror. Alternatively, a white light source can be used for illuminating the sample. A lens on a movable mount switches between confocal and wide field excitation as described later in this chapter. The collected photoluminescence light is filtered spatially by imaging onto a pinhole. Spectral filtering can be performed with a narrow bandpass interference filter. The light transmitted through these filters is directed on a CCD camera for imaging or to a grating spectrograph for spectral analysis. Finally a Hanbury Brown-Twiss correlation setup is used to measure the second-order coherence function.

3.1. Excitation laser

A frequency doubled, diode pumped neodymium:vanadate (Nd:YVO$_4$) laser (Coherent, Verdi) was used for cw excitation of the quantum dots. The laser wavelength is 532 nm. The output power can be varied between 10 mW and 10 W, but the laser was mainly operated in the low mW regime when used as the excitation light source. The output light is linearly polarized along the vertical axis and its specified linewidth at maximum power is 5 MHz.

The laser beam is collimated, and its diameter is extended to 1 cm with a confocal telescope. In this way, a beam waist broader than the entrance aperture of the microscope objective (0.75 mm) was achieved. This over-illumination is necessary to create a nearly diffraction limited spot size of the laser focus on the sample (see section 3.3).

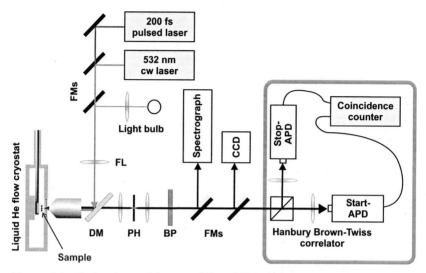

Figure 3.1.: *Basic scheme of the setup (FL and FMs: lens and mirrors on flip mounts, respectively, DM: dichroic mirror, PH: pinhole, BP: narrow bandpass filter, APDs: avalanche photodetectors).*

Pulsed excitation is done with a mode-locked titanium sapphire (Ti:Sa) laser (Coherent, Mira), pumped by the Nd:YVO$_4$ laser. The wavelength of this laser can be tuned between 750 nm and 930 nm. The laser is operated in the femtosecond mode and produces pulses at a repetition rate of 76 MHz. At the maximum pump power (10 W), optimum pulsing is achieved at an average output power of 1.4 W. With the measured pulse width[1] of 390 fs, this corresponds to a peak power of 50 kW.

Figure 3.2(a) shows an auto-correlation measurement[2] of the laser pulses. With transform limited pulses assumed, the fits yield a pulse width of 390 fs. Figure 3.2(b) displays a spectrum with a bandwidth of 1.38 nm, which corresponds to a similar pulse width of 380 fs. Although the pulses are broader than specified (200 fs) this does not affect the measurements in this work which have a minimal time resolution of 400 ps.

In order to obtain excitation above the exciton energy of the quantum dots, the laser is frequency-doubled by focussing onto a BBO (beta barium borate) nonlinear crystal.

---

[1]The spectral bandwidths and pulse durations are full widths at half of the maximum (FWHM) throughout this chapter.

[2]Here, a Michelson interferometer was used to measure the interference visibility as described in chapter 5.2.

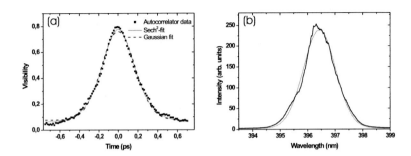

Figure 3.2.: *(a) Auto-correlation signal of the Ti:Sa laser pulses (dots). The lines represent two different fits for the data points. (b) The black line is a spectrum of the laser output after frequency doubling, the gray line is a Gaussian fit to the spectrum.*

The fundamental wavelength is suppressed by interference and colour glass filters. A subsequent telescope widens the beam diameter in the same way as it is done with the cw excitation beam path. The average power of the frequency-doubled light reaches 50 mW, which corresponds to a maximum intensity of 20 μW/μm^2, when illuminating a circular region with a diameter of 60 μm on the sample. When switching to confocal illumination, an intensity exceeding 10 mW/μm^2 can be achieved. The intensity of both laser beams can be attenuated continuously with a neutral density filter wheel.

3.2. Cryostat

The samples are cooled down to 4.2 K with a continuous-flow liquid Helium cryostat (CryoVac, Konti-Cryostat-Mikro). The cryostat interior was evacuated to achieve a pressure of 10^{-6}–10^{-7} mbar. A temperature sensor and a heating wire were used to actively control the temperature. This leads to temperature stabilization within 0.1 K at temperatures between 4 and 30 K. At higher temperatures, this stability decreases slightly. A two axes step motor inside the cryostat enables the horizontal positioning of the whole cooling finger within a range of 5 mm per axis and an accuracy of 0.1 μm. This is sufficient to reach every location on the sample with a precision below the microscope resolution. At a constant temperature, mechanical drifts in this cryostat are small enough for operation without realignment over one hour.

3.3. Microscope setup

The microscope setup is a crucial part of this experiment because here, the highest losses appear, when collecting single photons (see section 3.6). Moreover, the resolution of the microscope limits the ability to spatially separate two quantum dots located close together.

In a conventional microscope, the resolution can be defined by a point-like monochromatic object and its diffraction pattern in the image plane, the so-called "point spread function". In the case of a circular objective aperture, the image shows an Airy disc pattern. The Rayleigh criterium defines the resolution as the distance Δx of the central peak of the Airy disc to its first minimum [BW99]:

$$\Delta x = 0.61 \frac{\lambda}{\text{NA}}, \tag{3.1}$$

with wavelength λ. The numerical aperture is defined as NA$= n \sin \theta$, where n is the refractive index of the medium between sample and objective and θ is half the angular aperture of the objective.

Higher collection efficiency (see also section 3.6) and enhanced optical resolution can be obtained with a larger numerical aperture of the microscope objective. However, the use of the sample at low temperatures and inside a vacuum chamber prohibits the application of immersion oil and requires a minimum working distance.

A more sophisticated technique to further enhance the resolution and to suppress unwanted detection of stray light is Scanning Confocal Microscopy. The principle of confocal microscopy is depicted in figure 3.3(a). The illuminating light field is formed by a plane wave light source. In many setups, the plane wave approximation is achieved by focussing a laser onto a pinhole, which forms a point light source, and collimating it, again. In this way the microscope objective forms an image of the pinhole on its object plane. This gives a minimum spot size, limited by diffraction at the objective. Thus, only a small volume of the sample is illuminated. Then, the sample is imaged onto a second pinhole so that the excitation point source, the illuminated sample spot and the second pinhole form conjugated points, in other words, they are confocal. The light transmitted through the pinhole will be collected by a photodetector. In order to form an image of the sample, either the sample or the excitation beam has to be scanned. However, in this work, imaging of larger sample areas was performed with conventional microscopy, and one switched into the confocal mode when looking at single emitters.

The effect of the pinhole can be described by the three cases sketched in figures 3.3(c)–(e): Light emerging from the focal point of the microscope system will pass through the pinhole (c). However, light from outside the alignment axis (figure (d)) or away from the focal plane (e) will mostly or totally be blocked by the pinhole and does not contribute to the detected signal. As a result only fluorescence from the small

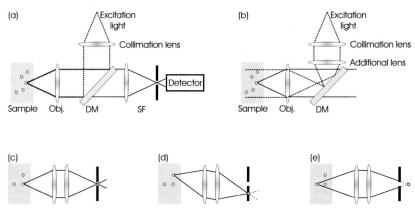

Figure 3.3.: *(a) Schematical setup of the confocal microscope. The excitation beam path is plotted dashed, the detection path with a solid line. Obj. denotes a microscope objective (represented by a single lens here), DM is a dichroic mirror and SF a spatial filter that consists of a lens and a pinhole in its focal plane. (b) Variation of the excitation beam path when an additional lens focusses the excitation light into the back focal plane of the objective for wide field illumination. (c)–(e) Principle of confocal imaging as described in the text.*

illuminated volume is detected. This strongly suppresses the background of the detected signal and additionally enables the acquisition of two-dimensional cuts through a three-dimensional object. The intensity distribution from a single emitter in the object plane of a confocal microscope is determined by the point spread functions of both the excitation point source and of the emitter itself. In the ideal case, this results in a distribution proportional to the square of the Airy pattern, and the resolution becomes:

$$\Delta x = 0.44 \frac{\lambda}{\mathrm{NA}}, \tag{3.2}$$

which exceeds the resolution of a conventional microscope, equation (3.1).

In this work, a microscope objective (Zeiss, $63\times$ Achroplan) was used, which combines a very high numerical aperture (NA = 0.75) with the required working distance (1.3 mm) for observations through the cryostat window. The microscope objective is corrected for abberations when using glass cover slips (in this case the cryostat window). A piezoelectric mount enables fine adjustment of the focal distance to the object. The excitation laser is collimated and expanded in a telescope arrangement to approach the field of a plane wave. A lens on a kinematic mount, which focusses the laser in the objective's back focal plane, can switch between confocal and wide-field illumination

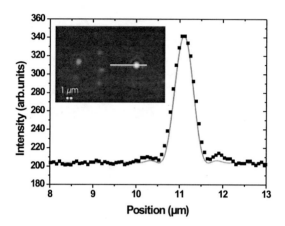

Figure 3.4.: *Intensity distribution of a single quantum dot (squares) along the white line, indicated in the inset. The gray curve is the expected Airy pattern of a point source with the experimental parameters NA=0.75 and $\lambda = 690$ nm.*

(sketched in figures 3.3(a) and (b)). A dichroic mirror with 570 nm marginal wavelength (below the quantum dot emission) pre-separates laser and collected light. Imaging onto the pinhole plane and collimation of transmitted light is done with two 100-mm lenses. The pinhole has a size of 75 μm and can be flipped out of the beam path for conventional microscopy images.

A CCD image of several quantum dots is shown in the inset of figure 3.4. It was obtained under wide field illumination and without the pinhole. The gauging of the distances on the CCD plane was derived from imaging a precision test grating (designed for Atomic Force Microscopy). As the size of a quantum dot is on the order of 10 nm, a single dot can be treated as a point source and its image can be used to determine the microscope resolution. A cross-section through one of the quantum dot images (squares in figure 3.4) has a width of $\Delta x_{cut} = 0.53$ μm and overlaps well with the expected Airy function (gray curve in figure 3.4) with a width of $\Delta x_{Airy} = 0.48$ μm.

The actual effective diameter of the circular pinhole used in the confocal mode is 1 μm which is larger than the measured possible resolution but gives a trade-off between high resolution and high transmission (leading to a high overall collection efficiency).

3.4. Spectrograph

Spectra are taken by using a 0.5-m grating spectrograph in combination with an intensified liquid-Nitrogen-cooled CCD camera (Roper Scientific, Acton 500i). With an entrance slit size of 100 μm and a 1800-mm^{-1} grating (blazed for 700 nm), the spectral resolution can be measured to 100 μeV by observing the Nd:YVO$_4$ laser linewidth (which is specified to 0.02 μeV). According to the specifications of the spectrograph with this grating, the resolution changes only insignificantly within the wavelength range of 400–800 nm that is used in this work.

3.5. Hanbury Brown-Twiss correlator

The Hanbury Brown-Twiss correlator was realized by splitting the incident light with a 50:50 beam splitter cube and focussing onto two silicon avalanche photodetectors (APDs), as shown in figure 3.1. In contrast to photo multiplier tubes (PMTs), APDs offer much higher detection efficiencies and substancially lower dark count rates. Nevertheless, PMTs become advantageous if a high time resolution is needed. In the wavelength regime of this work, the two APDs (Perkin-Elmer, SPCM) have a detection efficiency between 45% at 500 nm and up to 70% at 700 nm (see the gray line in the inset of figure 4.1(b)). The electrical dark count rates are 60 s^{-1} and 75 s^{-1}, respectively. A black box with only a small opening for the incoming light beam was built around the correlator to reduce count events from background light. Together with dimmed laboratory light conditions, an experimental dark count rate (including electrical and background counts) between 100 s^{-1} and 150 s^{-1} was achieved.

As reported by Kurtsiefer et al. [KZM$^+$01], it has been observed in this setup that secondary emission of light after photo-detection in Si-APDs gives rise to a cross talk between the two detectors in a Hanbury Brown-Twiss arrangement. Here, this manifests in a bunching peak artefact at around 1 ns away from the time origin. In order to suppress this peak (and to further reduce background light), small apertures were mounted directly in front of the detectors.

The correlation function was measured by taking the time differences between start and stop count events from the two detectors. This was achieved with time correlation electronics (PicoQuant TimeHarp), based on a time-to-amplitude converter and a subsequent digitalizer. The same electronics was used for the lifetime measurements in chapter 5.4.2. Time differences are collected in a histogram with 37-ps time bins.

By measuring the second-order coherence function of the femtosecond laser pulses (figure 3.5), an overall time resolution of 800 ps was observed. The major contribution to this value emerges from the time resolution of the two APDs (400 ps).

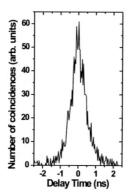

Figure 3.5.: *Second-order coherence function, measured on the fs pulsed laser.*

3.6. Overall photon generation efficiency

The total photon generation efficiency η_{tot} can be defined as the product of different partial efficiencies that are discussed below:

$$\eta_{\text{tot}} = \eta_{\text{exc}} \times \eta_{\text{rad}} \times \eta_{\text{coll}} \times T \times \eta_{\text{det}}. \tag{3.3}$$

Here, the excitation efficiency η_{exc} is the probability, that the quantum dot state is populated by an excitation pulse, whereas the probability for radiative decay is η_{rad}. The collection efficiency is η_{coll} and the transmission through the setup is T. Finally the detection efficiency is denoted by η_{det}.

On a first view, the excitation efficiency η_{exc} is just the probability of populating an electron and a hole in the lowest excited state of a quantum dot. At sufficiently high pump pulse power, the transition is saturated (figure 3.6) and the excitation efficiency tends to unity, as each laser pulse creates many charge carriers above the band gap, which are efficiently captured by the quantum dot.[3] This is valid at low temperatures. As the temperature increases, thermal escape of the carriers from the quantum dot also have to be considered and decrease η_{exc}.

---

[3]This is in contrast, for example, to resonantly excited single atoms, where no population inversion can be achieved.

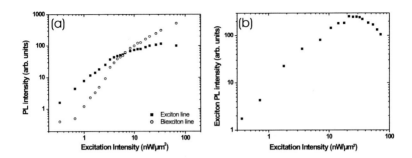

Figure 3.6.: *Photoluminescence (PL) intensity of the exciton (squares) and biexciton (open circles) spectral line of a single InP quantum dot under (a) pulsed and (b) cw excitation (exciton line only), respectively.*

Additionally, in cw excitation, especially the exciton state can be bleached effectively at high pump power. This happens when the rate to capture a second electron-hole pair and to form a biexciton state dominates the decay rate of the exciton. This results in a maximum in the exciton population versus the pump power (figure 3.6(b)). When further increasing the pump power, the multi-exciton states can bleach in a similar way.

In the measurements on single photon generation, a maximum excitation efficiency was not the only aspect. A clean spectrum with few and weak additional spectral lines was also important, so that the quantum dots were not always driven to saturation, leading to an excitation efficiency in the measurements of this work between 40% and 100%.

The probability η_{rad} that a once created exciton recombines radiatively, is limited by the presence of non-radiative decay channels such as defects in the vicinity of the quantum dot. Measurements on an InP quantum dot sample similar to one used in this work showed, that the exciton recombination is not dominated by non-radiative decay channels [ZPH+99]. Therefore, a radiative decay probability close to one is assumed.

The collection efficiency η_{coll} describes the probability that a once emitted photon is collected and depends on various factors. Self-assembled quantum dots are usually overgrown by a capping layer of material with high refractive index n. This leads to a small critical angle θ_{tir} for total internal reflection:

$$\theta_{\text{tir}} = \arcsin(1/n). \tag{3.4}$$

Light emitted at a larger angle is totally reflected at the semiconductor–air interface and

33

cannot leave the medium at all. For example, with the InP/GaInP sample used in this work, $n = 3.5$ and therefore maximum angle is $\theta_{\text{tir}} = 17°$. The fraction of light that can escape the material is then:

$$\eta_{\text{esc}} = \frac{\Omega_{\text{esc}}}{4\pi} = \frac{1}{2}\left(1 - \sqrt{1 - 1/n^2}\right), \tag{3.5}$$

where Ω_{esc} is the solid angle of light (inside the material) that leaves the sample. For $n = 3.5$, this yields $\eta_{\text{esc}} = 2\%$.

Additionally, the collection efficiency is limited by the collection solid angle of the microscope objective, given by its numerical aperture:

$$\Omega_{\text{NA}} = 2\pi\left(1 - \sqrt{1 - \text{NA}^2/n^2}\right). \tag{3.6}$$

A large refractive index further deteriorates the collection efficiency, as the light leaves the medium with a shallower angle. When assuming no further losses by partial reflection on the surface, the total collection efficiency for s- and p-polarized light, respectively, is given by [ZB02]:

$$\eta_{\text{coll,s}} = \frac{3}{16}\left(\frac{5}{3} - \frac{13\sqrt{1 - (\text{NA}/n)^2}}{8} - \frac{\cos(3\arcsin(\text{NA}/n))}{24}\right) \quad \text{and} \tag{3.7}$$

$$\eta_{\text{coll,p}} = \frac{3}{16}\left(1 - \frac{7\sqrt{1 - (\text{NA}/n)^2}}{8} - \frac{\cos(3\arcsin(\text{NA}/n))}{8}\right). \tag{3.8}$$

With NA= 0.75 and $n = 3.5$ these efficiencies are $\eta_{\text{coll,s}} \approx \eta_{\text{coll,p}} = 0.8\%$. When including partial reflections, this has to be further corrected to smaller values (see [ZB02] and table below).

Various techniques have been developed to exceed this limit, like coupling the emitter to micro-cavities [GSG+98, SPY01] and the usage of solid immersion lenses [ZB02]. In the InP quantum dot sample presented in chapter 4.1, an aluminium mirror is coated below the quantum dot layer that reflects part of the light emitted in the backward direction and nearly doubles the collection efficiency.

The transmission T through the optical setup is given by the various optical elements, mainly filters, and depends on the wavelength of the transmitted light. With the current setup, estimations based on specifications and measurements of the optical components[4] give $T = 55\%–61\%$.

The detection efficiency η_{det} of the APDs strongly depends on the wavelength of the detected light (see inset of figure 4.1(b)) and reaches a maximum above 70% at 700 nm.

---

[4]As the distinct samples used in this work emitted at different wavelengths, party different optical components were used.

The table below gives an overview of the different efficiencies for both samples used in the experiments (InP and CdSe). The InP quantum dots emit just at the maximum of the detection efficiency of the detectors. On the other hand, the CdSe system has a lower refraction index allowing the collection of light within a larger solid angle. The overall efficiency was measured by exciting the system with the pulsed laser and comparing the detector count rates with the laser repetition rate.

Quantum dot sample	InP/GaInP	CdSe/ZnSe
Wavelength	690 nm	510 nm
Refractive index	3.5	2.7
η_{exc}	between 40% and 1	
η_{rad}	≈ 1	
η_{coll}	0.5%	0.9%
T	55%	61%
η_{det}	70%	50%
η_{tot}	0.08–0.2%	0.11–0.27%
measured	up to 0.3%	up to 0.1%

Deviations between measured and expected efficiency have different reasons. One is the detailed surface structure that is unknown and not included in the above discussion and but which alters the collection efficiencies. Moreover, the transition lines are not always excited to saturation but to optimum single photon generation.

4. Single-Photon Generation with Single Quantum Dots

In this chapter the experimental realization of single photon generation using single quantum dots is described. Two different samples have been studied: One contained InP quantum dots in a GaInP matrix and is called the *InP sample*. The other sample consisted of CdSe quantum dots in ZnSe and is referred to as the *CdSe sample*.

4.1. InP quantum dots in GaInP

InP quantum dots in GaInP are a particularly interesting system for generating single photons for free beam experiments, as their emission wavelength around 690 nm allows the highest possible detection efficiency with Si APDs presently available. However, they show disadvantages in fibre-coupled applications because their losses in glass fibres are much higher than at infrared wavelengths (4 dB/km at 700 nm compared to 0.2 dB/km at 1.55 μm are typical values in glass fibres [Hec02]). On the other hand, infrared photo-detection suffers from low efficiency and a bad signal-to-noise ratio. Thus, single photons from InP quantum dots are particularly important for free-beam experiments.

4.1.1. Sample structure

InP quantum dots in GaInP have been studied at the single dot level and revealed a bimodal size distribution [CSP+94, PHP+03]. There are two types of dots: (I) fully developed dots with a typical size of 15 nm in height, 40 nm in width and with a photoluminescence emission centred at 750 nm, and (II) smaller dots with a height of less than 5 nm and a width of about 40 nm. These smaller dots emit around 690 nm and it is believed that they are prestages in quantum dot formation. The spectra of the fully developed dots turned out to have broad emission lines (> 1 meV) and consist of several peaks even for low excitation power [PHP+03]. Therefore, experiments in this work concentrate on the smaller quantum dots.

The sample used in this section was grown by Metal-Organic Vapour Phase Epitaxy (MOVPE)[1]. Figure 4.1(a) shows the structure of the sample. On a GaAs wafer a 300 nm

---

[1] The sample was provided by the group of Prof. W. Seifert from Lund University (Sweden).

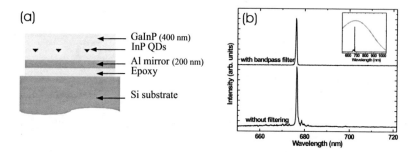

Figure 4.1.: *(a) Structure of the InP/GaInP sample. (b) Photoluminescence spectra taken on a single InP quantum dot under cw excitation and at a temperature of 10 K. The bottom spectrum was taken without filtering, the top spectrum was taken through a narrow bandpass filter. An offset was added for separating the graphs. The black line in the inset is a spectrum taken over a larger wavelength range, and the gray line shows the efficiency of the single photon detectors.*

thick GaInP layer was deposited, followed by 1.9 mono-layers of InP, which formed the quantum dots, and another 100 nm of GaInP. The density of dots emitting around 690 nm was estimated to be about 10^8 cm^{-2} by imaging through a narrow bandpass filter. In order to increase light extraction efficiency, a 200 nm thick Al layer was deposited on top of the sample to form a mirror. The sample was then glued upside down with epoxy onto a Si substrate and the GaAs substrate was removed using a selective wet etch of H_2O_2:NH_4OH:H_2O with a ratio of 1:1:10. For the purpose of increasing the light extraction efficiency, the use of a metallic mirror is preferable to a distributed Bragg reflector (DBR), as metal mirrors reflect strongly for all angles, resulting in a larger integrated reflectivity if a point-like emitter is assumed.

4.1.2. Single-photon generation

Figure 4.1(b) shows a spectrum of a single InP quantum dot at 10 K. The lower curve is an unfiltered spectrum. The excitation power density was adjusted to have only one dominant emission line. By measuring the intensity of this spectral line, a linear dependency on the excitation power was observed, indicating an exciton transition. Additional emission lines appear with increasing laser power. A spectrum taken over a wider wavelength range is displayed in the inset of figure 4.1(b) and shows that all

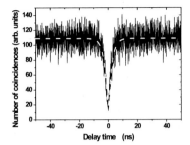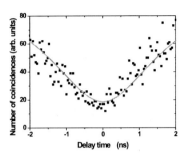

Figure 4.2.: *Measurement of the $g^{(2)}$-function under continuous excitation. The gray curve is the expected correlation function for an ideal single photon source but limited time resolution. The right graph is a magnification of the dip in the left plot.*

the emission within a very broad wavelength range originates from the dot under study. The inset also shows the detection efficiency of the APDs with its maximum right at the emission wavelength of the quantum dot. When placing a narrow 1-nm bandpass filter into the beam path, only light from a single transition of this quantum dot is transmitted (upper curve of figure 4.1(b)).

Figure 4.2 shows a correlation function (see equation (2.2)) measured on the exciton spectral line of this dot performed under continuous excitation [ZAP⁺03]. The total count rate[2] was 1.1×10^5 counts per second. The dashed gray line in this figure is the calculated correlation function obtained by taking into account the limited time resolution of the Hanbury Brown-Twiss setup. This calculated function is modelled as a convolution of the expected shape of the (ideal) correlation function $g^{(2)}(\tau) = 1 - \exp(-\gamma\tau)$ and a Gaussian distribution with a width of the system's time resolution of 800 ps. $1/\gamma$ is the time scale of the anti-bunching dip and depends on both the transition lifetime and the excitation time scale. This time scale is used as a fit parameter here. A zoom into the region around the origin of the left graph is given in the right graph of figure 4.2. The excellent agreement between the calculations and the measurement indicates that the quantum dot device generates single photons and that the minimum dip value of 5% (relative to the value at high time differences) is mostly due to the limited time resolution. The characteristic timescale $1/\gamma$ of the anti-bunching dip is fitted to 2.3 ns.

In figure 4.3 correlation measurements under pulsed excitation are displayed [ZAP⁺03].

---

[2]This is the sum over the count rates from both detectors.

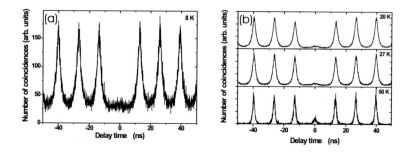

Figure 4.3.: *Second-order coherence function measured on a single InP quantum dot under pulsed excitation: (a) at 8 K and (b) at 20 to 50 K.*

The total count rate was 4.4×10^4 counts per second for the measurement in figure 4.3(a). It is observed that the peak at zero delay time is vanishing almost completely. This amounts to single photon generation on demand: Upon each laser pulse which creates an exciton, one and only one photon is produced. A measurement result of this type is the basis of the succeeding experiments in the next chapters because this justifies conclusions to individual quantum dots and one can exclude the possibility to observe several quantum dots at once, leading to ensemble averaging of the results. Additionally, for the quantum cryptography experiment described in chapter 7.2, only a normalized area of the zeroth peak clearly below 0.5 ensures secure transmission of the encryption key.

The graphs in figure 4.3(b) show measurements taken at higher temperatures. When increasing the temperature, the emission intensity of the quantum dot decreases, which can be attributed to thermal carrier escape. Moreover, broadening of the spectral lines due to phonon interactions (see [BLS+01, MZ04] and chapter 5) leads to an increased incoherent background when other spectral lines start to overlap with the filter transmission window. Both of these two effects deteriorate the single photon generation quality with respect to photon generation efficiency and photon statistics. However, up to 27 K, the zeroth peak is still almost completely suppressed. With increasing temperature this peak slowly starts to grow, but even at 50 K it has an area still below 0.5 relative to the higher order peaks, which indicates that still a single quantum dot transition dominates the emission.

4.2. CdSe quantum dots in ZnSe

In this section the generation of single photons with single Stranski-Krastanov grown CdSe quantum dots in a ZnSe matrix is described. Self-assembled II-VI quantum dots offer the advantage of stability, narrow emission spectrum and sub-nanosecond lifetime. Compared with III-V dots, the shorter lifetime of 300 ps [FHL+01, BWS+99] allows the generation of single photons on demand with a better time accuracy and a higher repetition rate of up to 3 GHz. Also, when studying coherent effects, such as quantum beats or entanglement generation, the probability of dephasing during this shorter lifetime is reduced. ZnSe has a refractive index of 2.7 [UOO+93] and total internal reflection takes place only for an incident angle larger than 22°, leading to an increased extraction efficiency from the sample by a factor of 2 compared to GaInP and GaAs embedded emitters.

4.2.1. Sample structure

The sample was grown by Molecular-Beam Epitaxy (MBE) [RLH98][3]. Previous studies by Atomic Force Microscopy (AFM) and Transmission Electron Microscopy (TEM) revealed the shape and sizes of these dots with heights of 1.6 nm and base lengths shorter than 10 nm [LRG+02]. AFM studies of similar samples revealed a dot density of 10^{11} cm^{-2}, which means an average dot density smaller than the microscope resolution. To perform single-quantum-dot spectroscopy, mesas with lateral extensions of 100 nm were etched, now containing only few quantum dots each [FHL+01].

4.2.2. Single photon generation

The photoluminescence of an ensemble of quantum dots shows an inhomogeneously broadened spectrum centred around 515 nm with a width of 15 nm. The sample is naturally n-type doped [Yao85]. As a result donor electrons may be captured by the quantum dots and, together with optically excited electron–hole pairs, form negatively charged excitons (trions). In contrast to the experiments with InP quantum dots in this chapter the pulsed excitation laser was tuned to 459 nm, so that no free charge carriers were excited in the ZnSe bandgap (440 nm). Spectral filtering was performed with a 3 nm bandpass filter.

Figure 4.4(a) shows spectra under different excitation powers, taken on a mesa that contains only a small number of quantum dots. The operating temperature was 6 K and the integration time was kept constant. The two spectral lines of interest are labelled as X$^-$ and XX$^-$. The inset of this figure shows the intensity dependency of these two lines (taken from Lorentzian fits) from the excitation power density. From the linear

---

[3]This sample was provided by the group of Prof. F. Henneberger at Humboldt-Universität zu Berlin.

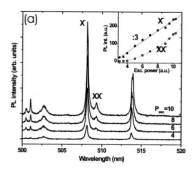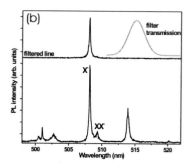

Figure 4.4.: *(a) Spectra from a single mesa under pulsed excitation at different excitation power densities. Power values are in units of 4 µW/µm². Offsets are added to the different curves for clarity reasons. The inset shows the intensities of the two spectral lines labelled as X⁻ and XX⁻ versus the excitation power. The solid and dotted lines are a linear and a quadratic fit, respectively. The intensity of the X⁻ data (filled squares) is scaled down by a factor of 3 for clarity. (b) Spectra before (bottom curve) and after (top curve) the bandpass filter at the experimental conditions for the correlation measurements described in the text. The gray curve shows the filter transmission, which was tuned to the spectral lines by tilting the filter. Again offsets are added to the curves.*

and quadratic behaviour, the lines could be verified as exciton and biexciton transitions, respectively. The energy spacing of 6 meV (1.2 nm) indicates the quantum dot to be negatively charged [AHF⁺02], in contrast to neutral quantum dots that typically show an energy spacing around 20 meV [BWS⁺99]. In figure 4.4(b), spectra before and after the bandpass filter are depicted as well as the bandpass transmission at the excitation power used for correlation measurements.

The intensity correlation measurement on the filtered transition is displayed in figure 4.5(a) and shows a pronounced anti-bunching, again [AZB⁺03]. Due to the short lifetime in this quantum dot system, the peak widths are dominated by the time uncertainty of the Hanbury Brown-Twiss setup. The area under the central peak at $\tau = 0$ reaches a normalized value of 0.28. The peak areas were normalized by integrating the four nearest peaks and setting their average to unity. This relatively large peak area, compared to the results in the previous chapter, is raised by a contribution of the emission from the charged biexciton and other neighbouring lines to the detected light. This can be estimated in the following way: If we assume that light transmitted through the bandpass filter is in a statistical mixture of a single- and a two-photon state, we can write the

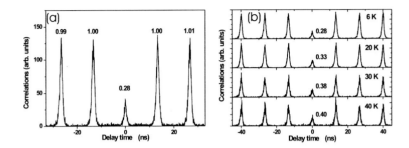

Figure 4.5.: *(a) Intensity correlation function measured on a single CdSe quantum dot at 6 K. The numbers next to the peaks indicate the normalized peak areas. (b) Correlation function taken on the same dot at different temperatures between 6 and 40 K. The numbers again indicate the peak area.*

density matrix as:

$$\hat{\rho} = (1 - \alpha)|1\rangle\langle1| + \alpha|2\rangle\langle2|. \tag{4.1}$$

Here α is the probability for having two photons in the state. Contributions from more than two photons are neglected, since, on average, much less than two photons per pulse are measured. Photon losses would result in an additional vacuum term $|0\rangle\langle0|$ in this density matrix but do not affect the results of the correlation measurements and are thus also ignored in this consideration. For such a state, the second-order coherence function at time zero is:

$$g^{(2)}(0) = \frac{2\alpha}{(1 + \alpha)^2} . \tag{4.2}$$

Thus, the measured value of $g^{(2)}(0) = 0.28$ gives an admixture of "foreign photons", including photons from the close biexciton line, of $\alpha = 22\%$.

The integration time in this measurement was 35 minutes and the photon count rate was 1.4×10^4 counts per second at an excitation power density of 40 $\mu W/\mu m^2$. As the exciton transition was not yet saturated, stronger excitation gives even higher count rates. However, this happens to the prize of an increased contribution of neighbouring spectral lines to the correlation function. In the experiment count rates up to 10^5 per second and central peaks with a normalized area still below 0.5 were observed.

Equivalently to the InP quantum dots in the previous section, the number of emitted photons decreases drastically at higher temperatures, so that the signal-to-noise ratio at the photon detectors drops as well. Together with a broadening of the spectral lines, this

43

results in a further deterioration of the photon statistics. On this sample, anti-bunching below a central-peak value of 0.5 was observed up to 40 K (see figure 4.5(b)).

4.3. Re-excitation processes in quantum dot devices

In this section, an anomaly is discussed that is sometimes observed in pulsed second-order correlation measurements on the photoluminescence of the InP quantum dot sample [AZB04]. In figure 4.6, measurements on three different quantum dots are displayed, where this effect is differently pronounced. In all these cases, the excitation light source was the 200-fs pulsed laser at 400 nm. The temperature ranged between 8 K and 15 K. In figure 4.6(a) a correlation function is shown, which is similar to the ordinary correlation functions expected from single photon generation as measured also in figure 4.3. This measurement shows a clear anti-bunching, with a minimum at $\tau = 0$, which indicates that, due to the spectral and spatial filtering process, only one quantum dot transition is observed. However, within the order of 1 ns, long before the next excitation laser pulse impinges, there is a slight increase (bunching) of correlations. Figure 4.6(b) shows a quantum dot for which this effect is even more pronounced and with a higher bunching time scale (2.5 ns). The higher order peaks are relatively broad and start to overlap, which leads to an increased offset of the whole correlation function. Note, that this is not due to unwanted background light, as the coincidences still significantly drop below this offset at $\tau = 0$. The very extreme case, where this broadening is strong enough to completely wash out the correlation peaks, was observed, too. This can be seen in figure 4.6(c). Still, there is a strong anti-bunching effect. In all these measurements, the proper mode-locking of the pulsed laser was checked. These observations are not unique but occur in a similar way on several quantum dots on this sample.

We attribute this effect to a re-excitation process that takes place in the quantum dot sample and that has the following origin: The optical excitation above the barrier band gap creates a large number of electrons and holes in the conduction and valence band, respectively. These electrons and holes can be captured in the quantum dot and relax into the exciton ground states, where they perform the decay transition that is investigated. However, some charge carriers may also be captured for a certain time in charge traps in the barrier material close to the quantum dot. After the initially excited exciton has decayed, further electrons and holes can diffuse from the traps into the dot and repopulate it. An additional signature of the charge traps is the inhomogeneous spectral linewidth broadening that is observed on these quantum dots (see chapter 5).

To gain further insight into this recapture process, different correlation measurements on one single quantum dot have been acquired with increasing excitation power. Figure 4.7 shows these data. The different power densities are achieved by inserting a neutral filter with varying optical density which does not affect the laser pulse shape. Anti-bunching dips show that only a single transition was observed. However, for increasing

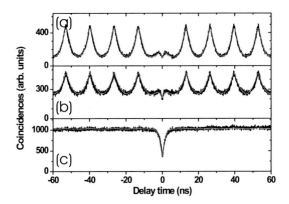

Figure 4.6.: *Correlation measurements on three different InP quantum dots (black curves). The sample temperatures were 15 K for measurement (a) and 8 K for measurements (b) and (c). All three graphs were obtained under pulsed excitation. The gray curves are results from Monte Carlo simulations.*

pump power, a bunching shoulder starts to grow close to the central minimum and the time scale of the anti-bunching dip decreases (see figure 4.7(b) and the shift of the guiding lines, indicating the bunching shoulders in (a)). In the measurement with an intensity of 100 nW/μm^2, the narrow anti-bunching dip at $\tau = 0$ is completely washed out by the finite time resolution of the detector.

A similar series of experiments with constant excitation power but varying sample temperature also gives a slight dependence of the re-excitation time, which decreases by a factor of 2 when increasing the temperature from 10 K to 40 K. This can be explained with an increase of the diffusion constant of the carriers in the wetting layer and barrier material and agrees also quantitatively with studies on the exciton diffusion constant in quantum wells by Oberhauser et al. [OPH$^+$93].

It has to be mentioned, that in spite of such re-excitation, the quantum dot still emits only one photon *at a time*. This is evident from the correlation function dropping below 1 at zero delay time which is thus still showing anti-bunching, whereas the bunching shoulder only appears at around 1 ns. However, this effect is an obstacle when a triggered single-photon source with emission on demand is needed, as it prevents the quantum dot to emit only one photon *per excitation pulse*. This is necessary in many single-

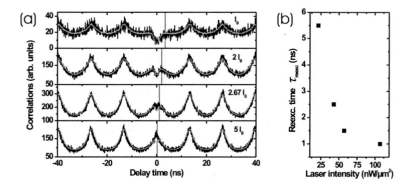

Figure 4.7.: *(a) Correlation measurements on an exciton spectral line for different excitation intensities. The black and gray curves are the measured and simulated correlation functions, respectively. The numbers denote the excitation intensities in units of $I_0 = 20$ nW/μm^2. The vertical guiding lines mark the shoulders of the bunching peak. (b) Re-excitation time scale τ_{reexc} versus excitation power density, obtained by the simulations.*

photon applications, where the photon emission needs to be strictly correlated to the excitation pulse, whereas photons in between the pulses need to be strongly suppressed. Nevertheless, as the re-excitation is very differently pronounced among the quantum dots, the careful choice of an observed dot, which shows no bunching shoulder, eliminates this obstruction.

4.3.1. Monte Carlo simulations

To support the theory of the re-excitation process, Monte Carlo simulations have been performed. These simulations are based on a common model for the decay cascade in single quantum dots [SFV+04]: A quantum dot can be excited by capturing electron-hole pairs that occupy the exciton, biexciton, etc. states. If the excitation energy is below the band gap of wetting layer and barrier material, these pairs are created directly in the quantum dot. If excited above the band gap, the charge carriers are created in the quantum dot environment and are subsequently captured in pairs or as single charges by the quantum dot. In this way, both charged and uncharged states can be formed (dashed arrows in figure 4.8, indicating excitation). In contrast to the excitation

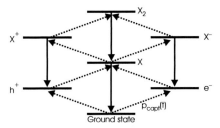

Figure 4.8.: *Model used for the Monte Carlo simulations. Solid arrows indicate radiative decay, dashed arrows excitation through electron and hole capture, with a probability* $p_{\text{capt}}(t) \propto \exp(-t/\tau_{\text{reexc}})$ *starting at the arrival time of the laser pulse.*

process, the decay of a (bi-) exciton does not change the quantum dot charging as it is a recombination of one electron and one hole (solid arrows in figure 4.8). This mechanism is responsible for long-term correlation effects that are reported in [SFV$^+$04] and also addressed in chapter 7. As the exciton lifetime in quantum dots is much larger (in this sample around 1 ns) than both the relaxation time (several 10 ps [VMR$^+$96]) and the laser pulse width (4000 fs), an *immediate* excitation at the impinging time of the laser pulses describes the ground-state exciton formation in good approximation.

For the simulations presented here, the model introduced in [SFV$^+$04] was extended by the fact that charges can be trapped for some time in the barrier material. The density of charges is assumed to be proportional to the excitation power and to decay exponentially with some time constant τ_{reexc}. Finally, the rate with which charge carriers are captured is assumed to be proportional to the charge carrier density, thus:

$$p_{\text{capt}}(t) = p_{\text{capt}}(0) \exp(-t/\tau_{\text{reexc}}),\qquad (4.3)$$

where t is the time elapsed since the latest excitation laser pulse.

The gray lines in figures 4.6 and 4.7(a) show simulated correlation functions. In the three simulations of figure 4.6, the simulation parameters describing the re-excitation process were fitted to the experimental data. For the simulations in figure 4.7(a), the initial capture rate $p_{\text{capt}}(0)$ is kept proportional to the excitation power used in the measurements. The re-excitation time scale τ_{reexc} was varied to fit the experimental data and the simulated curves. This time scale was found to depend strongly on the excitation intensity as displayed in figure 4.7(b). An additional offset had to be introduced in the simulated data to compensate for incoherent background light, especially at higher excitation powers. The following table summarizes the used parameters:

Simulation time step	10 ps
Simulated detector resolution noise amplitude	400 ps
Laser repetition time	13.2 ns
Probability that a captured charge is an electron (hole)	0.5 (0.5)
Transition rate	2 GHz
Initial capture rate $p_{\mathrm{capt}}(0)$	$0.4 \text{ GHz} \times \frac{\text{(Excitation intensity)}}{\text{nW}/\mu\text{m}^2}$
Re-excitation time scale τ_{reexc}	1–6 ns (see figure 4.7(b))

The detector resolution was introduced by adding Gaussian noise with an amplitude given in the table. As exciton and biexciton lifetimes were not known in detail, these two parameters were set equal for simplicity. Also the probability that a neutral dot captures an electron or a hole is kept equal for both charges.

As the simulations agree very well with the measured results, they support the proposed model for the re-excitation process. This model was also tested on a set of correlation functions measured on other quantum dots and also revealed a good overlap between experimental and numerical data. The parameters $p_{\mathrm{capt}}(0)$ and τ_{reexc} are found to vary strongly from dot to dot.

5. Fourier Spectroscopy with Single Photons

In the previous chapter, measurements of the photon statistics, in form of the second-order coherence function $g^{(2)}(\tau)$, were performed on the photoluminescence of single quantum dots. These measurements show the single photon character of the dot emission. However, many applications, such as quantum computation using single-photon qubits [KLM01] and quantum teleportation [BPM+97] additionally require consecutive photons to have identical wave-packets. There are two important parameters to describe this property of the light field: the spontaneous emission lifetime and the longitudinal coherence length[1] s_{coh} or equivalently the coherence time $\tau_{coh} = s_{coh}/c$. When comparing these two quantities, conclusions can be made about how distinguishable photon wave-packets of successive emissions are.

Coherence length and spectral linewidth, which are conjugate quantities, were measured on the emission of self-assembled quantum dots using high resolution spectroscopy [BF02, BKM+01], four-wave mixing [BLH01, BLS+01] and coherent spectroscopy in the time domain [BEG+98, TSN+00] (for a review see also [Mas02]). Another technique which provides direct information about the line shape, linewidth and coherence time of an emitter is Fourier spectroscopy. Fourier spectroscopy is an ideal method for single-photon spectroscopy since it combines high precision and a possible robust setup with very low photon losses. Measurements using this method for single quantum emitter spectroscopy have been reported on InAs/GaAs quantum dots in a pillar microcavity [SFV+02] and on mesa structures [KCV+02].

In this chapter, Fourier spectroscopy on single InP quantum dots using a Michelson interferometer is described. Photon correlation measurements on the same investigated spectral lines prove the observation of individual quantum dot transitions. From a more fundamental point of view, this combines measurements of the wave-like and the particle-like nature of light. Time resolved photoluminescence studies complement these measurements and allow the comparison of lifetime and coherence time.

---

[1]In the following s_{coh} is simply referred to as the *coherence length*.

5.1. Spectral properties of single emitters

The lifetime of an excited state (here the exciton and biexciton states) is determined by spontaneous decay. One distinguishes two decay types: Radiative decay, where the energy difference is carried away by an electro-magnetic wave, and non-radiative recombination, like absorption and emission of phonons or population of other states as in Auger processes.

In the simple case of an excited state with a single transition to the ground state, the population of the excited state is $n(t) \propto \exp(-t/\tau_{\mathrm{sp}})$ where the spontaneous emission lifetime is related to the transition rate: $\tau_{\mathrm{sp}} = 1/(2\gamma)$. In the case of several decay channels, the individual transition rates add to an effective rate $\gamma = \sum_i \gamma_i$. The electric field of the emitted wave is of the form:

$$E(t) = E_0 e^{-\gamma t} e^{i\omega_0 t}. \tag{5.1}$$

In an ensemble measurement one distinguishes two types of spectral broadening. If all ensemble members undergo the same transition, the spectrum is called *homogenously broadened*. This is especially the case when one deals with a set of identical emitters in identical physical environments. The spectrum is then given by the Fourier transform of the field and results in a Lorentz shaped curve:

$$S_{\mathrm{hom}}(\omega) = |\tilde{E}(\omega)|^2 \propto \frac{\gamma}{(\omega - \omega_0)^2 + \gamma^2}. \tag{5.2}$$

One sees that the spectral linewidth is $\Delta\omega = \gamma = 1/(2\tau_{\mathrm{sp}})$. It is called the *natural* or *homogeneous linewidth*.

If varying physical conditions induce a jitter of central frequency ω_0 for each transition, the resulting spectrum is called *inhomogeneously broadened*. Examples of this process are the Doppler broadened spectrum of an atomic cloud and the spectrum of a set of quantum dots with varying size and exciton transition energies. This concept can also be expanded to the *time averaged* spectrum of a single emitter. The resulting spectrum then is a convolution of the homogenous spectrum with a certain distribution function $g(\omega, \Gamma)$ around the central wavelength, characterized by its width Γ:

$$S_{\mathrm{inhom}}(\omega) = \int_{-\infty}^{\infty} d\omega'\, S_{\mathrm{hom}}(\omega - \omega') g(\omega', \Gamma). \tag{5.3}$$

Such a spectrum is determined by a Lorentzian shape if $\gamma \gg \Gamma$, whereas for $\gamma \ll \Gamma$ it follows the inhomogeneous distribution. As a result, the homogeneous transition rate γ forms a lower bound for the linewidth of the spectrum.

In the case of single-photon states, the spectrum has to be naturally described by ensemble measurements: Although such a state can have a well-defined spatio-temporal wave function with a unique Fourier transform (which can be identified as its spectrum), any attempt to measure the spectrum of only one photon, for example in a grating spectrograph, will lead to either one or no detection event, which allows no conclusion about the overall spectrum.

In a homogeneously broadened spectrum, the single-photon can be described by a pure number state or Fock state. Photons from different emission cycles of such a system are indistinguishable, which is necessary for the application in various quantum devices. This is especially the case in experiments including two-photon interference [HOM87], such as in linear optical quantum computation [KLM01]. However, if the inhomogeneous broadening dominates, the photon is in a statistical mixture of different modes and has to be described by density matrix formalism. The ratio of twice the lifetime to the coherence time is commonly used to quantize this difference. Generally it is $2\tau_{sp}/\tau_{coh} \geq 1$ and in the special case of a Fourier limited spectrum: $2\tau_{sp}/\tau_{coh} = 1$.

5.2. Michelson interferometry

The Michelson interferometer was invented in 1880 by A. Michelson and improved together with E. Morley [Mic81, MM87]. Originally, it was used to search for the evidence of Earth's motion through ether, but soon it also became a tool for interferometric spectroscopy. In its basic form, the Michelson interferometer consists of a beam splitter that divides the incoming light field into two interferometer arms of different lengths. Thereupon, the two beams are reflected back and merged at the beam splitter as depicted in figure 5.1. The two partial beams show interference depending on their coherence properties.

The following derivations are based on the formalism of Mandel and Wolf [MW95] and are adapted to the experimental details of this work. If the incoming field is denoted by $E_{in}(t)$, the field leaving the interferometer has the intensity[2]:

$$I = \left\langle |E_{out}(t)|^2 \right\rangle_T = I_0 - \left\langle \text{Re} \left(E_{in}^*(t - \frac{2L_1}{c}) E_{in}(t - \frac{2L_2}{c}) \right) \right\rangle_T. \tag{5.4}$$

The time delays are given by the two interferometer arm lengths $L_{1,2}$, respectively, as defined in figure 5.1. The minus sign is caused by an internal reflection at the beam splitter. By detecting the outcoming field with a detector whose time resolution T is much larger than the time scale of the oscillating electric field, the time-averaged intensity is measured. This is denoted by $\langle x(t) \rangle_T \equiv \int_{-T/2}^{T/2} x(t + t') dt'$. The intensity contribution of each interferometer arm is $I_0 = \langle |E_{in}(t + 2L_{1,2}/c)|^2 \rangle_T$. They are equal in

---

[2]The spatial dependence is omitted in this description, as the interferometer correlates the electric field at different times and therefore spatial variations do not play a direct role. However, if not aligned properly or for a strongly diverging beam they do inflict the measurement results.

both arms of a balanced interferometer, as it is the case in the experiments presented here. Due to averaging, they become time independent. This is still valid in the single photon limit, if the detector integration time is much bigger than the average arrival time distance between two adjacent photons. In this case, additional ensemble averaging has to be performed in equation (5.4).

Now we assume a quasi-monochromatic incoming light field of the form $E_{in}(t) = E_0(t) \exp(i\omega_0 t)$ with the central frequency ω_0 and a linewidth of $\Delta\omega$. When performing the time averaging on a larger time scale than the coherence time $T \gg \tau_{coh}$, the intensity is found to be:

$$I = I_0 - \text{Re} \left[2\pi \int_{-\infty}^{\infty} d\omega \underbrace{\left| \tilde{E}_0(\omega) \right|^2 e^{i\omega s/c}}_{S(\omega + \omega_0)} \right] \cos(\omega_0 s/c), \qquad (5.5)$$

where the interferometer path length difference $s = 2|L_1 - L_2|$ is introduced. $\tilde{E}_0(\omega)$ denotes the Fourier transform of $E_0(t)$. According to the Wiener-Khintchine theorem, the expression in the square brackets is the Fourier transform of the optical spectrum $S(\omega + \omega_0)$ (with the maximum shifted to the origin), which also relates the coherence length and the linewidth [MW95].

In order to directly obtain the Fourier transform of the spectrum a convenient quantity, the visibility $V = (I_{max} - I_{min})/(I_{max} + I_{min})$, is calculated. It is bound between 0 and 1 according to zero and maximum interference contrast, respectively. I_{max} and I_{min} are the maximum and minimum intensity of the interference pattern, respectively. These maxima and minima are dominated by the cosine term in equation (5.5), so that the visibility is proportional to the Fourier transform of the spectrum:

$$V(s) = \frac{2\pi}{I_0} \text{Re} \left[\int_{-\infty}^{\infty} d\omega \, S(\omega + \omega_0) e^{i\omega s/c} \right]. \qquad (5.6)$$

The resolution of a Michelson interferometer is given by its maximum path length difference. In order to resolve two wavelengths λ and $\lambda + \Delta\lambda$, it is necessary to observe at least one period of the resulting beat signal. The first minimum of the beat signal occurs if the interference maximum of the mth order of wavelength λ coincides with the minimum of wavelength $\lambda + \Delta\lambda$, that is if $m\lambda = (m + 1/2)(\lambda + \Delta\lambda)$. This leads to the following condition for the minimum path length difference:

$$s_{min} = \frac{\lambda(\lambda + \Delta\lambda)}{2\Delta\lambda} \approx \frac{\lambda^2}{2\Delta\lambda}. \qquad (5.7)$$

A similar condition arises when defining the resolution with respect to the linewidth: For instance a Lorentzian spectrum $S(\omega) \propto 1/((\omega - \omega_0)^2 + \Delta\omega^2)$ with linewidth $\Delta\omega$ shows an exponentially decaying visibility $V(s) \propto \exp(-|s|\Delta\omega/c)$. In order to provide at least the distance at which the visibility drops to $1/e$, a minimum path length difference of $s_{min} = c/\Delta\omega = \lambda^2/(2\pi\Delta\lambda)$ is required.

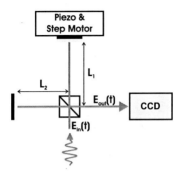

Figure 5.1.: *Sketch of the Michelson interferometer.*

5.3. Interferometer setup

The Michelson interferometer was realized with a fixed mirror in one arm and a movable mirror in the other arm. This second mirror is mounted on a piezoelectric translator with a range of 80 μm. The coarse position is adjusted with a step motor translation stage with a 5 mm range as depicted in figure 5.1. In order to minimize mechanical drifts, the components were mounted on a whole massive aluminum block. With this, typically, drifts within $\lambda/10$ occur on a time scale of 1 min. The whole interferometer is mounted in front of the CCD camera of the setup sketched in figure 3.1.

Using a CCD camera has two major advantages over point detectors such as APDs: First, the direct observation of the two partial images of the interferometer arms allows an easy and quick alignment of the mirrors. Additionally it is possible to perform measurements in parallel on several dots in a given area under exactly the same experimental conditions (see also figures 5.2 and 5.6). The slightly lower detection efficiency (50% at 700 nm) with respect to an APD has to be compensated with a higher integration time, which is between 100 and 500 ms per image.

Figure 5.2(a) shows a sequence of three CCD images of the interferometer output at different mirror positions, where the intensity of the visible dots varies between destructive and constructive interference. In order to acquire the full interference pattern, a movie was taken while driving a linear voltage ramp on the piezoelectric translator. As the camera readout software also saves the acquisition time of each frame, a direct assignment of each picture to the piezo position can be made. A software was written to extract the intensity of a given dot dependent on the piezo position. Figure 5.2(b) shows the intensity distribution of the quantum dot marked in figure 5.2(a). The background was subtracted from the signal by measuring the noise level on a nearby region of the CCD with the same area where no dot was present. A sinusoidal fit was then used to

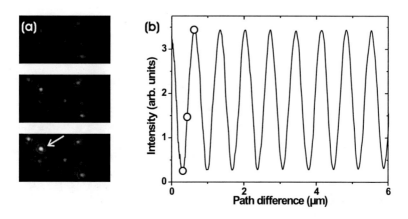

Figure 5.2.: *(a) Images taken through the Michelson interferometer showing several InP quantum dots with varying intensity between destructive and constructive interference. (b) Single quantum dot photoluminescence intensity for the dot marked by an arrow in (a) as a function of the path length difference. The circles indicate the three positions where the images were taken.*

extract the visibility value for each measurement.

For testing the interferometer, several light sources of known specifications were used. In figure 3.2 the visibility trace of the pulsed Ti:Sa laser was compared with its spectrum which shows good agreement. Figure 5.3 shows the visibility dependence of a white light source transmitted through a narrow bandpass filter and of the cw Nd:YVO$_4$ laser. The specified coherence length of the laser of 60 m is much larger than the maximum travel range of the interferometer. Only a slight decrease in the measured visibility due to the divergence of the laser beam was observed. The spectral bandwidth[3] of the filter obtained from the Fourier spectrum is (1.2 ± 0.2) nm and matches the bandwidth taken directly from the spectrum $((1.12 \pm 0.02)$ nm) and from the specifications (1.2 nm). The visibility shape of the bandpass filter transmission is also of importance for the interpretation of the Fourier spectra in the next section.

---

[3]Again all linewidths are full widths at half of the maximum (FWHM).

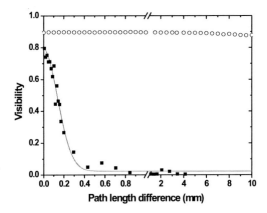

Figure 5.3.: *Fourier spectra of the cw Nd:YVO$_4$ laser (open circles) and of a white light source transmitted through the narrow bandpass filter (filled squares). The gray line is a Gaussian fit to the filter data. Note the change of scale at the path length difference of 1 mm.*

5.4. Fourier spectroscopy on exciton and biexciton transitions

In this section, Fourier spectroscopy measurements on exciton and biexciton transitions of single quantum dots on the InP sample are described and discussed (see also reference [ZAB04]). Spectra at different excitation powers were taken to identify these transitions by means of their linear and quadratic dependence, respectively. Before each interferometry measurement, photon correlations were measured to confirm the anti-bunched nature of the quantum dot emission. Then, Fourier spectroscopy was performed individually on each line by filtering with the bandpass filter.

Figure 5.4(a) shows the visibility of the exciton emission of a single dot measured over a 8-mm scan at a temperature of 6 K. The inset shows the second-order coherence function measured on the filtered exciton emission under the same conditions and on the same dot. The normalized height of the peak at zero time delay is 33%, which is clearly below 0.5 and demonstrates that the emission originates from a single dot. The Fourier spectrum of the exciton transition consists of two regimes. Similar to measurements on InGaAs self-assembled quantum dots [BLS+01] this behaviour corresponds to a narrow

55

emission line (zero-phonon line) on top of broader shoulders. In the measurements presented here, the shortest coherence length (or the broadest width of the shoulder in the corresponding spectrum) is limited by the bandpass filter.

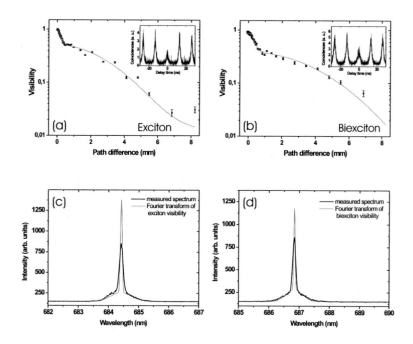

Figure 5.4.: *Visibility dependence of (a) the exciton and (b) the biexciton line of a single quantum dot, respectively. The gray curves are double-Gaussian fits. The insets show the corresponding second-order coherence functions of each line obtained at the same experimental conditions. The gray curves in (c) and (d) are the Fourier transforms of the fits in (a) and (b), respectively. The black curves are the spectra of the exciton and biexciton lines after the bandpass filter.*

A fit based on two Gaussian functions was used to extract the coherence lengths and to reconstruct the linewidths. The choice of the Gaussian function is motivated with hindsight by the observation of spectral diffusion. However, here, this function plays the role of a mere model which is used to extract the spectral bandwidths. The extracted coherence length[4] of the exciton zero-phonon line is (2.5 ± 0.1) mm, yielding an exciton emission linewidth of (187 ± 7) μeV.

The visibility was also measured on the biexciton emission of the same dot under the same conditions and is displayed in figure 5.4(b). Here again, the inset shows a photon correlation measurement demonstrating single-photon emission from the biexciton recombination with a normalized peak size of 46%. The coherence length of the biexciton is (3.3 ± 0.1) mm which corresponds to a linewidth of (142 ± 4) μeV.

The short scale coherence lengths are (220 ± 30) μm for the measurement on the exciton and (280 ± 10) μm for the biexciton. This amounts to spectral shoulders with widths of (4.2 ± 0.6) meV and (1.67 ± 0.06) meV, respectively.

The graphs in figures 5.4(c) and (d) display the Fourier transforms of the fits to the two visibility distributions (gray curves). They are confronted with the directly measured spectra of exciton and biexciton line after the bandpass filter (black curves). Fourier transforms and spectra were scaled so that the areas under each curve become equal. Central wavelength and offset of the Fourier transforms were adjusted manually to overlap with the spectra. Due to the lower resolution of spectrograph the zero-phonon lines in the direct spectra appear broader and weaker compared to the corresponding lines measured by Fourier spectroscopy. For the broad spectral shoulders, there is a good overlap between spectra and Fourier transform, as their measurement is not resolution limited.

5.4.1. Temperature dependence

In a next step, the dependence of the zero-phonon spectral lines on the temperature was observed. Figure 5.5(a) shows Fourier spectra of the exciton line of another dot at a temperature varying between 7 K and 40 K. Again, the anti-bunched nature of the lines was checked as can be seen in figure 5.5(d). The measurements show an increasing linewidth on both timescales with increasing temperature. The extracted temperature-dependent linewidths are presented in figure 5.5(b). In the same graph, the results from the biexciton emission of another dot is plotted as well. For exciton and biexciton emission from the same dot, typically similar linewidths were retrieved, as discussed for the results displayed in figures 5.4.

The linewidth remains at a lower limit up to a temperature of 20 K and then starts to enhance with increasing temperature. The broadening of the linewidth of the zero-phonon line with increasing temperature has been also observed in photoluminescence spectra of single II-VI type quantum dots [BKM+01] and in four-wave-mixing experiments on InGaAs quantum dot ensembles [BLS+01]. It is attributed to phonon assisted transitions to higher energy exciton states which competes with radiative recombination

---

[4]Here, the standard deviation of the Gaussian envelope was taken.

of the exciton at higher temperature. The appearance of the broad spectral shoulder on top of the narrow spectral line has its origin in the phonon coupling, as well [MZ04].

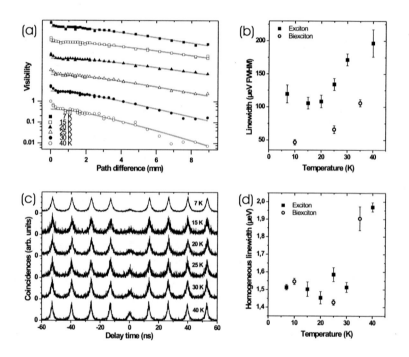

Figure 5.5.: *(a) The visibility traces of an exciton transition of a single quantum dot at different temperatures. An offset was added to each trace for clarity. (b) The linewidth extracted from the fits in (a) versus temperature. In the same graph the results from a similar set of measurements on a biexciton line are plotted. (c) Second order coherence function obtained at the same experimental conditions as for the measurements in (a). (d) Homogeneous linewidth versus temperature, extracted from the measured lifetime of each state.*

5.4.2. Comparison of coherence time and emission lifetime

In order to get information about the purity of the emitted single photon states, the obtained coherence time has to be compared with the lifetime of the exciton states.

The lifetime can already be determined by the measurement of the second-order coherence function under pulsed excitation: There the width of the higher correlation peaks of a single emitter is already dependent on the emitter's lifetime. This is due to the fact, that the pulsed laser immediately excites the emitter, whereupon the signal at each photodetector is proportional to the population probability of the excited state. This results in the higher order peaks being described by the self-convolution of the population probability function:

$$g_i^{(2)}(\tau) \propto \int_{-\infty}^{\infty} dt\, n(\tau + t)\, n(t) \quad \propto \quad e^{-|\tau|/\tau_{\text{sp}}}, \tag{5.8}$$

where $g_i^{(2)}(\tau)$ indicates the correlation function in the vicinity of the i-th peak. The last equality holds for the case of an exponential decay $n(t) \propto \theta(t)\exp(-t/\tau_{\text{sp}})$, where the step function $\theta(t)$ assures the initialization by the laser pulse.

A more direct way to obtain the lifetime is to measure the time-resolved evolution of the photoluminescence signal. This can be achieved by a modification of the Hanbury Brown-Twiss correlation setup. Instead of acquiring the correlations between two single photon detectors, now the time difference between a fast photo diode detecting an excerpt of the excitation laser and the detection events of one single photodetector is measured.

The lifetimes of the exciton and biexciton transitions used for figure 5.4 turned out to be 1.03 ns and 1.34 ns, respectively. This corresponds to a homogeneous linewidth of $\Delta E = h\gamma = h/(2\tau_{\text{sp}}) = 2.01~\mu\text{eV}$ for the exciton and $\Delta E = 1.54~\mu\text{eV}$ for the biexciton transition.

Figure 5.5(d) shows the homogeneous linewidths taken from lifetime measurements and measured at the same transitions and temperatures as the data in figure 5.5(b). It can be seen that the homogeneous linewidth remains constant up to approximately 30 K. At higher temperature, other, non-radiative decay channels open, leading to a reduction of the lifetime and accordingly to a homogeneous linewidth broadening. This coincides with an observed reduction of the photoluminescence intensity.

A striking observation is the fact that the individual spectral lines are inhomogeneously broadened by a factor of $2\tau_{\text{sp}}/\tau_{\text{coh}} \approx 50\ldots100$ relative to the expected homogeneous linewidth. This shows that decoherence processes are at work in these quantum dots even at low temperatures. It has been speculated that Stark shifts induced by charging effects and fluctuating charge traps in the barrier material of InP quantum dots create a spectral diffusion, which leads to an effective broadening of the spectral line [BWH+00]. The re-excitation process described in chapter 4.3 is evidence of these charge traps as

59

well. Although no blinking or spectral diffusion was detected on a time scale longer than 10 ms on the investigated dots, it cannot be ruled out that such spectral diffusion happens on a much shorter time scale, leading to the observed linewidths. The statistical process of the spectral diffusion justifies the usage of the Gaussian fit rather than an exponential fit in the visibility measurements.

Due to the incoherent broadening, the emitted photons cannot be treated as pure states, but have to be described by a probability distribution (see equation 5.3). At a first glance this broadening excludes the utilization of this quantum dot system for experiments including two-photon interference, where Fourier-limited, indistinguishable photons are required. But it has turned out that it is sufficient that *successive* photons remain identical within the range of their homogeneous linewidth. This would be still the case if spectral diffusion happens only on an intermediate timescale, larger than the repetition time of the excitation laser. For instance Santori et al. [SFV+02] could have observed a pronounced Hong-Ou-Mandel dip despite an inhomogeneous broadening of up to $2\tau_{\mathrm{sp}}/\tau_{\mathrm{coh}} = 7$.

A possible improvement to overcome this linewidth broadening is to excite the quantum dot quasi-resonantly into a higher exciton state but below the wetting layer and barrier material electronic states. In this way, at low temperatures, the creation of charges in the dot vicinity and the according interaction with the exciton states will be suppressed. Moreover, it has been shown that the photon linewidth was reduced when the quantum dot was excited close to resonance, as the energy deposited in the system is completely converted into light, leaving no excess energy that can excite phonons [KVC+02]. Resonant excitation requires a fine tuning of the excitation laser wavelength to the quantum dot absorption line together with a good stray light filtering to efficiently separate excitation and photoluminescence light.

The narrower biexciton linewidth is a repeatedly observed feature on this sample. Generally, the contrary case is expected, as the biexciton with its two electron–hole pairs exhibits twice as much recombination channels as the exciton. An explanation for this behaviour is presented in reference [Fli04]: The biexciton energy is twice the exciton energy reduced by the biexciton binding energy $E_{XX} = 2E_X - E_{XX}^B$. The jitter of the exciton and biexciton transitions both originate from a Stark shift due to the fluctuating local electric field. As they see the same field, in a good approximation the two transitions shift in the same direction albeit with different amplitudes. This suggests a linear relation between the biexciton binding energy and the exciton energy $E_{XX}^B = QE_X$ with $Q > 0$. Now the energy of the biexciton transition is determined by the difference between the biexciton and exciton energy $E_{XX\to X} = E_{XX} - E_X = (1 - Q)E_X$, whereas the exciton decays to the ground state with a transition energy of E_X itself. For the energy fluctuations, this means $\Delta E_{XX\to X} = (1 - Q)\Delta E_X$ which is indeed smaller than the exciton fluctuations.

5.4.3. Parallel acquisition of the coherence length

As mentioned before, imaging on a CCD array also allows to perform measurements in parallel on several *single* dots in a given area under exactly the same experimental conditions. The results of a parallel coherence length measurement on the emission from eight dots are shown in figure 5.6. The visibility dependencies of these eight dots were extracted from a single mirror scan. Variations in the single-photon coherence length between 1.2 mm and 3.2 mm (linewidth: 300–800 μeV) are observed. Because these different results are obtained simultaneously at the same energy and under the same excitation conditions, the variations in coherence length reveal intrinsic differences in the dephasing mechanisms from quantum dot to quantum dot.

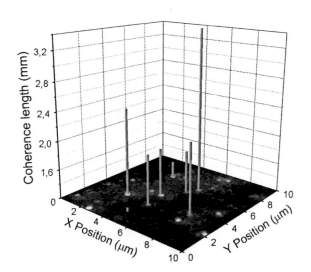

Figure 5.6.: *Result of a parallel coherence length measurement performed on several dots at a temperature of 6 K. The bars indicate the coherence length of the single photons emitted from each corresponding dot. The x–y plane displays a CCD camera image.*

5.5. Simultaneous measurement of particle and wave properties

The wave–particle duality of matter lies at the heart of quantum mechanics. With respect to light, the wave-like behaviour is perceived as being classical and the particle aspect as being nonclassical, while for massive microscopic objects, like neutrons and atoms, the opposite holds. The occurrence of an interference pattern is a manifestation of the wave-nature of matter.

Already in 1909, soon after the introduction of the concept of light particles (the term "photon" was created not until 1926, originally to describe chemical intra-atomic forces [Lam95]), it was observed experimentally that there is no deviation from the classically predicted interference pattern if a double-slit interference experiment is performed with very weak light, even if the intensity is so small that on average only a single photon is present inside the apparatus [Tay09]. Later, this observation was accounted for theoretically by quantum mechanics and was confirmed by more precise experiments [JN58, RSS69]. There exists an exact correspondence between the interference of the quantum probability amplitudes for each single photon to travel along either path in an interferometer, on the one hand, and the interference of the classical field strengths in the different paths, on the other hand. Therefore, the outcome of any first-order interference experiment can be obtained by describing light as a classical electromagnetic wave, independent of the statistical distribution of the incident photons. In these early experiments, both wave- and particle-aspect were observed in a single experiment when, with very small light intensities from a classical source, the double-slit interference pattern was gradually built up by registering more and more spots on the screen. A small non-vanishing probability remained that such a spot is not caused by a single photon, but by two photons arriving at the same time, though. According to the principle of complementarity, it is impossible to simultaneously observe interference and to detect which path each photon travelled in the interferometer.

Later, Grangier et al. [GRA86] performed a series of experiments with single photons from atomic decays. In a first step, they showed the single-photon character of the atomic emission by observing the corresponding anti-bunched behaviour of the intensity correlation function. In a second step, they inserted the photons into a Mach-Zehnder interferometer and observed an interference pattern with varying path difference, a feature that displays the wave nature of light. Braig et al. [BZK+03] implemented a similar experiment using a diamond defect centre as emitter, where they observed single-photon statistics after detecting interference in a Michelson interferometer. From this fundamental point of view, the Michelson interference experiments in this chapter and the observation of anti-bunching in the previous chapter also form measurements of both the first and second order coherence function and show wave- and particle-like outcomes.

In this section, an experiment is described, where these two experimental techniques

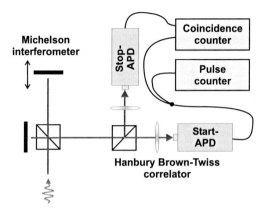

Figure 5.7.: *Experimental setup for the simultaneous Michelson and Hanbury Brown-Twiss experiment.*

are combined in a single step for *simultaneous* observation of interference and anti-bunching of the quantum dot fluorescence [AHS+04]. For this, Michelson interferometer and Hanbury Brown-Twiss correlator were set together as displayed in figure 5.7. Each photon emitted by the quantum dot first interferes with itself at the beam splitter of the Michelson interferometer. The exit of the interferometer directly enters the Hanbury Brown-Twiss setup where the photons are detected by one of the APDs.

The output pulses of the detectors can be exploited in two ways: First, when looking for coincidences between pulses from start and stop APD, anti-bunching is observed, revealing the particle nature of light. The only effect from the change between constructive and destructive interference on this measurement is an overall change of the coincidence rate, independent of the delay time between start and stop events. Since the latter is short compared to the time scale of the arm length variation in the Michelson interferometer, the un-normalized second-order coherence function just changes by a constant factor. Second, one can count the detection pulses of only one APD, using pulse counting electronics that determines the count rate at a given time. In this case, the temporal interference pattern, a wave-feature of light, will be observable directly. Since the detector produces a classical electrical pulse that can, after detection of a photon, be easily split into two parts, it is also possible to perform these two measurements simultaneously.

Figure 5.8 displays the results of such a combined measurement when exciting a quantum dot with a pulsed ((a) and (b)) and a continuous laser ((c) and (d)), respectively. In these measurements the piezo, controlling one Michelson arm, was driven by a triangular

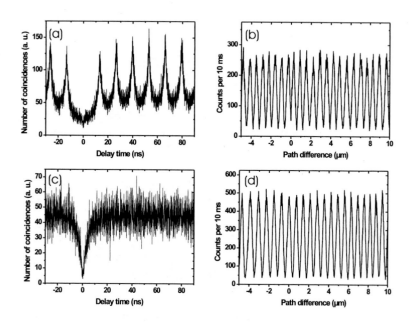

Figure 5.8.: *(a) and (b) measured correlation function and interference pattern, respectively, for a single quantum dot with pulsed excitation. (c) and (d) the same, but with continuous excitation.*

voltage signal corresponding to an amplitude of 20 μm (approximately 29 wavelengths) in the interferometer path difference, at a modulation rate of 10 mHz. In figures 5.8(a) and (c), the autocorrelation functions of the two measurements are plotted, expressed by the number of coincidences, integrated over 2 hours (pulsed) and 1 hour (cw). The nonclassical anti-bunching effect is clearly visible since the number of coincidences exhibits a pronounced minimum at zero time delay. In contrast, figures 5.8(b) and (c) depict the single-detector count rate, at an integration time of 10 ms, in dependence of the path difference in the interferometer, showing the expected first-order interference pattern, that reveals the wave-like nature of the emitted single-photon radiation.

The described combination of the two experimental techniques, Michelson interferometer and Hanbury Brown-Twiss correlation setup, forms an extension to the experiments of Grangier et al. [GRA86], as one and the same photon contributes to both the measured interference pattern and the anti-bunched correlation function. In this sense the described experiment is similar to the classic experiment of Taylor [Tay09] and to the experiments described in references [JN58, RSS69], but gives an unequivocal evidence of the particle nature of light: Instead of using weak light fields with classical photon number statistics (in this case a super-Poissonian photon number distribution), the anti-bunching effect shows that the quantum dot photoluminescence represents number states, that can only be described within the frame of quantum mechanics. In a similar work, Höffges et al. [HBE$^+$97] simultaneously performed heterodyne and photon correlation measurements in the resonance fluorescence of a single ion.

6. Multi-Photon Cascades in Single Quantum Dots

The potential of single semiconductor quantum dots as emitters in photonic devices is not only the generation of *single photons* on demand. Although it has not been realized so far, quantum dots are also promising candidates for the generation of entangled photon pairs. It was proposed [BSP+00] to make use of polarization correlations in the biexciton–exciton cascade, as described in the introduction of this work. But as will be demonstrated in chapter 7, even without entanglement formation, multi-photon cascades find an application in quantum communication experiments.

This chapter focusses on the investigation of intensity cross-correlations between several different quantum dot transitions on the InP quantum dot sample. Exciton–biexciton cross-correlation measurements, similar to those described here, have also been reported on InAs quantum dots [RMG+01, MRM+01, SFP+02] and CdSe quantum dots [USM+03]. Such experiments serve several purposes: First, they are an important tool for identifying the nature of the investigated spectral lines, such as resulting from an exciton, biexciton, triexciton or emerging from the same or different quantum dots. Second, they give information about the different decay and excitation time scales in multi-photon cascades. Finally, polarization resolved cross-correlations form a first step towards the observation of entangled photon pairs.

6.1. Setup for cross-correlation measurements

In order to detect correlations between different transitions, a variant of the second-order coherence function is considered. The *cross-correlation function* is defined in a similar way as in equation 2.2, but with the intensity operators assigned to different field modes α and β:

$$g_{\alpha\beta}^{(2)}(\tau) = \frac{\left\langle : \hat{I}_\alpha(t)\hat{I}_\beta(t+\tau) : \right\rangle}{\langle I_\alpha(t)\rangle\langle I_\beta(t)\rangle}. \tag{6.1}$$

In the experiments described here, the two modes represent spectral lines of two quantum dot transitions. In order to distinguish this cross-correlation from the second-order coherence function, the latter is also referred to as the *auto-correlation function* in this chapter.

The experiment is performed by spectral filtering of the two spectral lines for each photodetector in the Hanbury Brown-Twiss system individually. Here, this was realized by placing narrow bandpass filters directly in front of each APD, as sketched in figure 6.1. In this way, in general, the resulting correlation function will show an asymmetry with respect to the time origin, as start and stop detection events now arise from different processes and a change in sign of the time axis accords to an effective exchange of start and stop detector. In contrast to the measurements in chapter 4.1, narrower bandpass filters with 0.5 nm width were used.

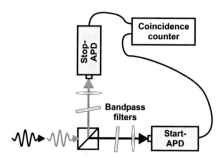

Figure 6.1.: *Modified Hanbury Brown-Twiss setup for the cross-correlation experiments. The different colours of the two beams represent two spectral lines, filtered by the bandpass filters.*

6.2. Cross-correlation measurements

The measurements that are described here were performed on the InP quantum dot sample (in this chapter referred to as sample A). Additionally a similar sample, with the same growth parameters, was used. On this second sample, a gold film with 460-nm circular apertures and a coordinate system was coated. These apertures expose only one or a few quantum dots and result in an additional reduction of the observed dot density and thus to an improved signal-to-background ratio. In this chapter this sample is denoted as sample B.

To get a first idea of the origin of the different spectral lines of a dot and to find the appropriate excitation intensity, first their different scaling with the excitation intensity was investigated. Figure 6.2 shows the photoluminescence spectra of a quantum dot from sample B, taken under various excitation intensities. All experiments on this dot were performed at 8 K. In the following figures, the reference excitation power density P_0 was kept constant at 1 nW/μm^2. The spectral behaviour in figure 6.2 is typical in

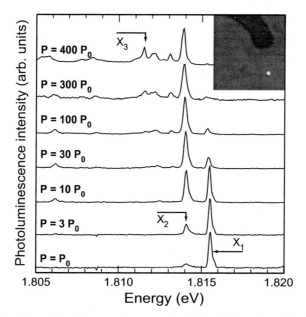

Figure 6.2.: *Power dependent spectroscopy of a single InP quantum dot showing the lines used in the correlation measurements. The excitation intensity is given as a multiple of $P_0 = 1 \ nW/\mu m^2$. Lines X_1, X_2 and X_3 are assigned in the text. The inset is a 40 μm × 40 μm photoluminescence image of sample B using a bandpass filter centred at 1.815 eV, which shows an aperture that contains a single dot emitting at this energy and a part of the coordinate system defined on the sample.*

terms of line spacing and power dependence, for an InP dot emitting in this energy range [PHP+03]. At a low excitation power density, a single sharp emission line at 686.3 nm (1.8155 eV) is present in the spectrum (X_1). As the excitation power is increased, a second line (X_2) appears about 0.6 nm (1.5 meV) below the exciton emission. When further increasing the excitation power, additional lines appear. The integrated photoluminescence intensity of X_1 increases linearly with the excitation intensity, whereas X_2 shows a quadratic dependence. This behaviour is a good indication of excitonic and biexcitonic emission, respectively. The lines appearing at high excitation power density, such as X_3, are attributed to a multi-exciton of higher complexity. Especially, X_3 is assumed to originate from a triexciton, as will be proven later.

For such a complex excitation as the triexciton, it is necessary to invoke additional states to the single-particle ground-states of the quantum dot. Figure 6.3(a) shows two of the possible triexciton decays. When varying the orientation of the electron and hole spins, one sees that the triexciton state is fourfold degenerated and four optical transitions to biexciton states are allowed [DGE+00]. In the transition I→II in figure 6.3, the energetically higher electron and hole recombine, leaving a ground-state biexciton. In the InP quantum dots system, the conduction-band level-splitting is on the order of 50 meV [PHP+03], and it should thus be possible to observe this transition at an energy roughly 50 meV above the exciton emission line. However, at this energy there is a strong competition with the wetting layer and the matrix material of the structure, hiding the interesting emission. In the other case of the triexciton decay, sketched in figure 6.3(a) as transition I→III, an electron and a hole in lower energy states recombine, which creates an *excited* biexciton which has an energy in the same range as the exciton and the biexciton transition. Altogether three combinations of this type of radiative triexciton decay are expected [DGE+00].

The simplified decay chain of a triexciton into an excited biexciton is sketched in figure 6.3(b). The triexciton X_3 recombines to an excited biexciton X_2^*, that rapidly relaxes to the biexciton ground state X_2, which in turn recombines via the exciton X_1 to the empty ground state G of the quantum dot. Auto-correlation measurements on each of the three radiative transitions are displayed in figure 6.3(c)–(e). An almost perfect anti-bunching dip for the exciton and the biexciton transition is observed, taking into account the response time of our photodetectors. The gray lines in the graphs 6.3 show the solutions from the rate model, described in section 6.2.1. For these solutions, only a normalization factor was used as a fit parameter, while the transition rates were kept constant. The offset of the anti-bunching dips in the predicted curves originates solely from an introduced convolution with a Gaussian to account for the detector resolution.

After the different spectral lines have been characterized and pre-identified, additional information can be gained by performing cross-correlation measurements between these emission lines. Figure 6.4(a) shows the cross-correlations of the exciton and biexciton line of that dot under different excitation power densities with cw excitation. A strong asymmetric behaviour is observed.

At positive time, when the detection of a biexciton photon starts the correlation measurement and the detection of an exciton photon stops it, photon bunching occurs, as here the detection of the starting biexciton photon projects the quantum dot to the exciton state, which has now an increased probability of recombining shortly after. On the other hand, if the correlation measurement is started by the exciton photon, which prepares the dot in the ground state, and stopped by the biexciton photon (negative times in figure 6.4(a)), a certain time is needed until the dot is re-excited. In this measurement, effectively the recycling time of the quantum dot is observed, which explains the strong anti-bunching for negative times. The population of the biexciton state is dependent on the laser power and the excitation time decreases when the laser power increases.

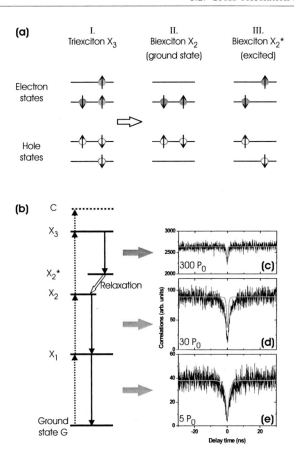

Figure 6.3.: *Illustration of the multi-exciton cascade in quantum dots. (a) Occupation of electron and hole states in the decay of the triexciton state X_3 to the biexciton ground state X_2 and to an excited state X_2^*. (b) Decay cascade model, used in the discussions and for the rate equation approach. Dashed arrows indicate excitation, solid arrows radiative decay and the open arrow a non-radiative relaxation. The dashed state C symbolizes an effective cut-off state as explained in the text. (c)–(e) Measured auto-correlation function of one of the triexciton transitions and of the biexciton and exciton spectral line, respectively. The gray lines in these graphs are the predictions from the rate model. The numbers are the excitation intensities as multiples of $P_0 = 1 \ nW/\mu m^2$.*

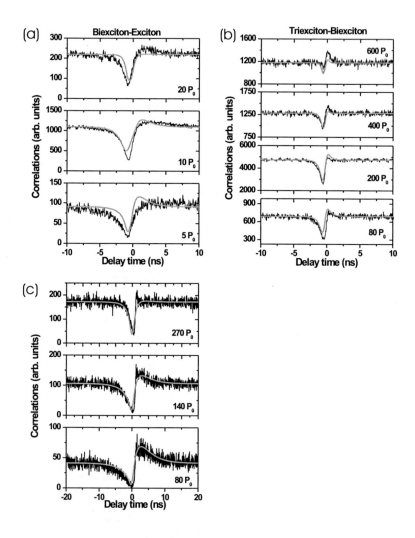

Figure 6.4.: *Measured cross-correlation functions (a) between the exciton and biexciton line and (b) between the biexciton and triexciton line of the quantum dot on sample B, that was also used for figures 6.2 and 6.3, at different excitation intensities ($P_0 = 1\ nW/\mu m^2$). (c) Exciton–biexciton cross-correlations of a dot on sample A. The gray curves are the predictions from the rate model.*

Similar measurements have also been performed on a single quantum dot on sample A. Figure 6.4(c) shows cross-correlations between its exciton and biexciton transition. While its time scales are similar to the data of the dot on sample B, here, the bunching peak appears more pronounced.

In the same way, the cross-correlation of the biexciton emission with the triexciton emission was measured (sample B). This is shown in figure 6.4(b). The behaviour is similar to the exciton–biexciton case but with different time scales apparent.

It was not possible to correlate the exciton with the triexciton, since the exciton is quenched at excitation intensities necessary to obtain a triexciton signal (see figure 6.2). Correlation measurements were also performed between the biexciton line and two additional lines, emitting at 1.812 eV and 1.813 eV, which become visible at high power ($400\,P_0$ in figure 6.2), but no (anti-) correlation was observed. These two lines are weak also at high excitation power density and the spectra seem to consist of many competing lines. It is also possible that some of these emission lines originate from another dot.

The presence of the combined bunching/anti-bunching shape is a unique hint for observing a decay cascade of two adjacent states. In contrast, the cross-correlation function of spectral lines of two independent transitions (for example from two quantum dots) would show no (anti-)correlations at all. It can be concluded that there is a three-photon cascaded emission from the triexciton via biexciton and exciton to the quantum dot ground state. Together with the information of the different scaling of the spectral lines with the excitation power, this justifies the previously made assignments to these lines.

6.2.1. Rate model

In order to support the interpretation of the obtained correlation data, the photon cascade was analysed using a common rate model [MRM+01, RMG+01, ZPH+99]. The rate equations correspond to the scheme shown in figure 6.3(b), where it was assumed that only two transition types are responsible for the dynamics of the excitonic states: spontaneous radiative decay and re-excitation with a rate proportional to the excitation power. As the excitation is performed above the quantum dot continuum, the relaxation of the charge carriers into the multi-exciton states, as well as the relaxation of the excited biexciton after the triexciton decay should also be taken into account. But as this process happens on a much faster time scale (several 10 ps [VMR+96]) than the state lifetimes ($\approx$ ns), it is neglected in this consideration. The according rate equation ansatz then reads:

$$\frac{d}{dt}\mathbf{n}(t) = \begin{pmatrix} -\tau_E^{-1} & \tau_1^{-1} & 0 & 0 & 0 \\ \tau_E^{-1} & -\tau_E^{-1} - \tau_1^{-1} & \tau_2^{-1} & 0 & 0 \\ 0 & \tau_E^{-1} & -\tau_E^{-1} - \tau_2^{-1} & \tau_3^{-1} & 0 \\ 0 & 0 & \tau_E^{-1} & -\tau_E^{-1} - \tau_3^{-1} & \tau_C^{-1} \\ 0 & 0 & 0 & \tau_E^{-1} & -\tau_C^{-1} \end{pmatrix} \mathbf{n}(t), \quad (6.2)$$

with $\qquad \mathbf{n}(t) \equiv (n_G(t), n_1(t), n_2(t), n_3(t), n_C(t))$.

Here n_1, n_2 and n_3 represent the populations of the exciton, biexciton and triexciton, respectively, with corresponding decay times τ_1, τ_2 and τ_3. n_G is the population of the empty ground state and τ_E^{-1} is the excitation rate. In order to truncate the ladder of states connected by rates in this model, an effective cut-off state with population n_C and life-time τ_C was introduced. This accounts for population and depopulation of all higher excited states via excitation and radiative decay, respectively.

An analytical solution of these equations can be obtained by diagonalizing the matrix in equation (6.2). In the eigen-basis, the solutions are simple exponentials and the general solution in the original basis is a sum of exponentials with different time constants:

$$n_i(t) = A_i + B_i e^{\lambda_1 t} + C_i e^{\lambda_2 t} + D_i e^{\lambda_3 t} + E_i e^{\lambda_4 t}, \qquad (6.3)$$

with $i \in \{G, 1, 2, 3, C\}$. λ_j are the (negative) eigen-values and $A_i, \ldots, E_i$ are constants, which have to be specified by initial conditions. As one eigen-value vanishes, the first term remains constant and can be used for normalization.

The initial conditions are defined by the transition that forms the *start* event in the Hanbury Brown-Twiss measurement, which prepares the quantum dot in the next lower state α, so that $n_\alpha(0) = 1$ and $n_{\gamma \neq \alpha}(0) = 0$. On the other hand, the detection of the photon from the *stop* transition dictates the shape of the cross-correlation function, as $g_{\alpha\beta}^{(2)}(t) \propto n_\beta(t)$. By this consideration it is clear that the cross-correlation function on the positive and negative side is described by two completely different functions with a possible discontinuity at $\tau = 0$. In the experiment (figure 6.4), this discontinuity is washed out, due to the finite time resolution of the detectors. Because of this smoothing of the experimental data, the minima in the graphs of figure 6.4 are shifted towards the anti-bunching side, as well.

The detailed expression of the eigen-values and the solution of equation (6.2) will not be given here, as they are very extensive expressions of the various transition rates, which do not offer direct physical insight. In order to account for the finite detector resolution the solution of the rate equation (6.3) was additionally convoluted with a Gaussian distribution $G(t) = (2\pi w^2)^{-1/2} \exp(-t^2/(2w^2))$ with a width $w = 800$ ps obtained in chapter 3.5.

For weak excitation $< 30P_0$, it can be seen in figure 6.2, that there is only a weak contribution of the triexciton line. In this case, a simplified model was used, which is analogue to equation (6.2), but with the biexciton as the highest cut-off state:

$$\frac{d}{dt}\mathbf{n}(t) = \begin{pmatrix} -\tau_E^{-1} & \tau_1^{-1} & 0 \\ \tau_E^{-1} & -\tau_E^{-1} - \tau_1^{-1} & \tau_2^{-1} \\ 0 & \tau_E^{-1} & -\tau_2^{-1} \end{pmatrix} \mathbf{n}(t), \qquad (6.4)$$

with $\qquad \mathbf{n}(t) = (n_G(t), n_1(t), n_2(t))$.

When comparing the solutions of the simple (6.4) and extended model (6.2) at an intermediate power $30\,P_0$ and with transition rates used for the curves in figures 6.3 and 6.4, a maximum deviation of 2.5% was found, which justifies this simplification. The solutions of this simpler model are derived and discussed in appendix A.

These two models (depending on the experimental excitation intensity) were used to describe the auto- and cross-correlation data in figures 6.3 and 6.4. The results are shown as gray lines in these graphs. The lifetimes of exciton and biexciton decay were taken from pulsed correlation measurements. This was not possible for the triexciton decay, as only weak anti-bunching was observed, so that the observation of the triexciton transition alone could not be guaranteed. Instead, it was fitted by optimizing the auto-correlation in figure 6.3(c). The derived value (see table below) in proportion to the exciton lifetime is in agreement with calculations made by Dekel et al. [DGE+00]. For the timescale of the introduced cut-off state, the same value was estimated. The excitation rate was chosen to optimally fit the correlation functions in these figures, but was kept linear to the experimental excitation power P throughout the graphs. In this way, apart from the one-time initialization of the experimentally inaccessible values τ_3, τ_C and τ_E, the normalization was the only real fit parameter in the graphs. No vertical offset was used to compensate the lift of the anti-bunching dips.

The same approach was used with the results from sample A in figure 6.4(c). Again, the lifetimes of the exciton and biexciton decay were obtained from pulsed measurements. For the modelling, now, solely the simplified model, equation (6.4), was used, in which the time scales τ_3 and τ_C were not needed. The excitation rate was estimated in the same way as before and, again, kept linear.

The table below shows the parameters used for the models in figures 6.3 and 6.4:

	Sample B, figs. 6.3 & 6.4(a)–(b)	Sample A, fig. 6.4(c)
Exciton lifetime τ_1	(1.2 ± 0.1) ns	(2.8 ± 0.8) ns
Biexciton lifetime τ_2	(1.1 ± 0.1) ns	(2.0 ± 0.1) ns
Triexciton lifetime τ_3	0.1 ns (fit)	(not used)
Cut-off lifetime τ_C	0.1 ns (fit)	(not used)
Excitation rate τ_E^{-1}	0.025 GHz $\times P/P_0$	0.0015 GHz $\times P/P_0$
Offset y_0	0	0

Apparently, the model describes the experimental data very good. However, at a closer look, some deviations can be discovered. Especially in the auto-correlation functions (figure 6.3), the predicted dip is systematically narrower compared to the measured. In the cross-correlation functions, the theoretical functions do not always exactly follow the experimental curves, as well. At a first glance, this can be explained by the fact that only two model parameters, τ_1 and τ_2, have been directly obtained by measurement. Additionally, the whole experimental data was acquired over a period of several hours, so that sample drifts may cause slight local variations in the excitation power, which deviate from the externally obtained value, given in the graphs. But on the other hand, the extended model, equation (6.2), with the unknown parameters τ_3 and τ_C was only used for the data where the triexciton transition was observed (i.e. for excitation powers $P > 30P_0$). Moreover, the excitation rate, the third unknown value, was kept linear to the experimental power over two orders of magnitude $(5 \ldots 600P_0)$ and still resulted in a good qualitative agreement. This suggests that the choice of parameters is reasonable and that there are further effects which are not covered by the model. One possible effect could be the presence of charged states, as used for the Monte Carlo simulations in chapter 4.3, but which have been neglected here, as it was beyond the scope of the simple rate equation model used here.

6.3. Polarization resolved cross-correlations

The previously shown auto- and cross-correlation measurements have shown the presence of multiple-photon cascades in single quantum dots. Under certain circumstances the adjacent photons of such a cascade can even be entangled in polarization [BSP+00]. But even if entanglement is abolished, polarization dependent correlations between exciton and biexciton decay are expected.

Figure 6.5 sketches the decay paths of a biexciton state in a quantum dot. Due to the Pauli exclusion principle, in the biexciton ground state, the electrons occupy both spin states with spin projection $S_{e,z} = \pm 1/2$. The same holds for the holes but with the heavy-hole spin $S_{hh,z} = \pm 3/2$. It follows that there are two radiative biexciton decay paths arriving in the two possible bright exciton states with opposing spin orientations.

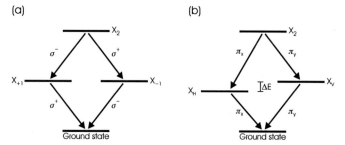

Figure 6.5.: *Scheme of the biexciton decay cascade, for (a) degenerated exciton states and (b) two exciton states with an energy separation ΔE. $\pi_{x/y}$ and $\sigma^\pm$ indicate linear and circular polarization of the emitted photon, respectively.*

When these two exciton states are degenerated (figure 6.5(a)), the bright exciton states with $S_{\text{total},z} = \pm 1$ are energy eigen-states, so that the first emitted, biexcitonic photon is in either right- (σ^+) or left-handed (σ^-) circular polarization, whereas the second, excitonic photon shows the opposite polarization. As long as no dephasing occurs during the exciton lifetime and the two decay paths remain indistinguishable, the two successive photons are entangled with respect to their polarization [BSP+00]. If the degeneracy is lifted, for example due to an anisotropic quantum dot environment, a mixing of the exciton levels occurs, which results in the emission of linearly polarized $(\pi_{x/y})$ photons (figure 6.5(b)) [BOS+02]. In this case, the different energies of the photons carry information about which path was taken, which destroys the entanglement. However, a correlation between the linear polarization is still expected, as the successive photons have the same polarization.

Experimentally, polarization correlation can be observed by additionally placing polarizers in front of the APDs in the cross-correlation setup of figure 6.1. With parallelly oriented linear polarizers (in the same basis as the dot emission) an asymmetric bunching/anti-bunching behaviour is expected again, when observing the exciton–biexciton transition, with the same arguments as for the unpolarized cross-correlation distributions. With orthogonally oriented polarizers however, the detection of a, say π_x-photon prepares the quantum dot in the — now unobserved — X_H state, and vice versa (see figure 6.5(b)). In this case, anti-correlations would be observed, instead. For the verification of entanglement one has to ensure that the transition from bunching to anti-bunching still occurs after basis transformation.

First polarization-dependent results, performed with a dot on sample A, are displayed in figure 6.6. In a preceding measurement the quantum dot emission was found to be slightly polarized with a degree of 30%. Apparently the symmetry in the figures 6.5 is broken, leading to one slightly preferred decay channel. An energy gap ΔE between the two exciton states was not observed, but is likely hidden by the inhomogeneous linewidth broadening. In the following, the axis defined by the weakly polarized light were used to define horizontal (H) and vertical (V) polarization. In the images in figure 6.6, the cross-correlation between exciton and biexciton with parallel $(HH$, figure (a)) and perpendicular $(HV$, figure (b)) polarization is displayed. First of all, it is noticeable that the predicted change from bunching to anti-bunching when switching from HH to HV configuration is not present. Still, when fitting an exponential of the form $f(t) = N(1 + \alpha \exp(-t/\tau))$ to the negative side of the data, a change of the bunching height from $\alpha_{HH} = 0.91 \pm 0.05$ to $\alpha_{HV} = 0.56 \pm 0.05$ occurs. Similar results were obtained when measuring in a basis rotated by 45°, with a change of relative bunching height from 0.93 to 0.54. Consequently, a measurement with mixed polarization bases (exciton: H, biexciton: +45°) yielded an intermediate height of 0.75. It seems that polarization correlations between photons from the exciton and biexciton transition are strongly disturbed. The excitation far above resonance and thus the large number of free charge carriers may be the main reason to obscure a clearer correlation as has been

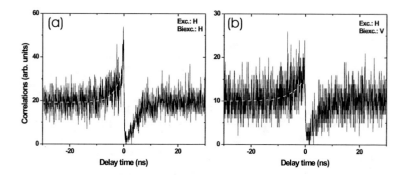

Figure 6.6.: *Cross-correlations between exciton and biexciton transition with varying polarizer orientations: (a) horizontal–horizontal and (b) horizontal–vertical. The dashed gray lines are exponential fits, as described in the text.*

observed from InAs and CdSe quantum dots [SFP+02, USM+03]. However, as a variation of the correlation heights between parallel and orthogonal polarization was observed in two orthogonal bases in a first preliminary result, InP quantum dots remain a promising system for observing at least partially entangled photon pairs.

7. Multiplexing on a Single-Photon Level and Demonstration of Quantum Cryptography

Among the common requirements for single photon sources, high efficiencies and high emission rates are a major priority, in order to raise the statistical significance of experimental outcomes or to enhance the bandwidth for quantum communication protocols. The overall efficiency can be improved by using passive optics such as integrated mirrors (like in the InP sample, see chapter 4.1) or solid immersion lenses to enhance the optical collection efficiency [ZB02], or by resonant techniques where the quantum emitters are embedded in micro-cavities [SPY01, GSG+98]. The latter method exploits the Purcell effect [Pur46] in order to enhance the emission rate in a certain well-defined resonant cavity mode. This Purcell effect can also substantially modify the overall spontaneous emission rate. For a single photon source which relies on the decay of an excited state the (modified) spontaneous lifetime determines the maximum photon generation rate.

In classical communication, multiplexing is a long used technique for increasing the transmission bandwidth. It is the transmission and retrieval of more than one signal through the same communication link (sketched in figure 7.1). This is usually accomplished by marking each signal with a physical label, such as the wavelength. At the receiver, the signals are identified by the use of filters tuned to the carrier frequencies [ST91]. Losses when merging and separating the signals can be compensated by amplification of the classical signal. For single-photon channels, the No-Cloning theorem [WZ82] prevents the amplification of qubit information, so that losses have to be kept minimal and effective separation of photons with different wavelengths is required. Moreover, a multi-colour single-photon source is needed to provide distinguishable photons.

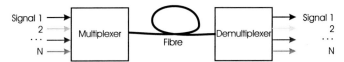

Figure 7.1.: *Transmission of N optical signals, here, distinguished by their wavelength, through the same fibre by the use of multiplexing.*

In this chapter, an interferometric technique to perform multiplexing on a single-photon level is demonstrated (see also reference [ARB04]). The biexciton–exciton cascade in quantum dots, described in chapter 6, provides an excellent source for the required photon pairs with well separated energies and strong correlation in the emission time. As a proof-of-principle, a quantum key distribution experiment using the BB84 protocol [BB84] was performed.

7.1. A single-photon add/drop filter

7.1.1. Principle of operation

In order to use several independent qubits in a single communication channel simultaneously, they have to be distinguishable in one physical property and a method is needed to merge and divide them at the sender and receiver side, respectively. For photons, a reasonable choice would be to distinguish them by their wavelength and to use their polarization to encode quantum information. A common way to separate light with different wavelengths is to use diffractive or refractive optics. However, these techniques are unfavourable, especially in inhomogeneously broadened systems, as for a sample of self-organized quantum dots. Here, the wavelength of the two photons as well as their wavelength difference may vary from dot to dot. When using diffractive and refractive optics, a complete realignment of the beam paths for each individual quantum dot under consideration would be required. Moreover, diffractive optics suffer from losses due to diffraction into different orders.

A superior method is to use interferometric techniques, like the one sketched in figure 7.2(a). Here, the two photons with different wavelengths λ and $\lambda + \Delta\lambda$ enter a Michelson interferometer with variable arm length. In contrast to the setup in figure 5.1, here, retro-reflector prisms were used to obtain a lateral shift between input and output beam. Due to the difference in wavelength the two photons undergo different interference conditions. As long as the path difference s between the two interferometer arms is significantly smaller than the coherence length s_{coh}, the probability to find a photon (with wavelength λ) at one of the two interferometer output ports is:

$$p_{1,2}(s,\lambda) = \frac{1}{2} \pm \frac{1}{2}\cos\left(2\pi s/\lambda\right) . \tag{7.1}$$

This is derived from equation 5.5 in the limit $s \ll s_{\text{coh}}$ and with the classical intensities replaced by probabilities as we assume single photon states. The signs $+$ and $-$ correspond to the interferometer output ports labelled as 1 and 2 in figure 7.2, respectively.

In order to illustrate how this can be used to separate the two photons, the interference pattern $p_1(s)$ for two different wavelengths is plotted in figure 7.2(b). For $s \approx 0$, initially, each wavelength shows the same interference pattern. But for increasing path difference

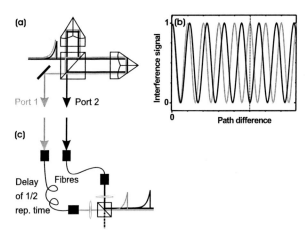

Figure 7.2.: *Sketch of the experimental setup: (a) Two Photons with different energies enter a Michelson interferometer that consists of a 50:50 beam splitter and two retro-reflectors. (b) Scheme of the intensity interference pattern at one interferometer output for two distinct wavelengths versus the path difference between the two interferometer arms. (c) Re-merging the separated photons: The two interferometer output ports are each coupled to an optical fibre, delayed relative to each other with half of the excitation laser repetition time and recombined at a beam splitter, again.*

they run out of phase and at a certain position (indicated by the dotted line in figure 7.2(b)) the two interference patterns are in opposite phase, that is, each photon interferes constructively at a different output port. The smallest path difference for which such a wavelength separation occurs is $s_0 = \lambda(\lambda + \Delta\lambda)/(2\Delta\lambda)$. As long as $s_0 \ll s_{coh}$, such a situation can always be achieved. Note that this condition simply reflects the spectral distinguishability of the two photons, which is a general limit for wavelength separation. When s_0 is in the order of or bigger than the coherence length, the interference visibility decreases and none or only poor photon separation is performed. In this case the setup basically acts as a 50:50 beam splitter.

It can be seen, that such an interferometric technique is highly insensitive to changes in wavelength difference, for example when changing the quantum emitter or in case of spectral drifts, as these effects can be compensated by simply correcting the interferometer arm lengths, which causes no change in the exiting, final beam direction. Moreover, as long as $s_0 \ll s_{coh}$, the main losses that occur in such a system are caused by partial back-reflection at the interfaces of the optical components, which can be strongly suppressed with appropriate anti-reflection coating.

Figure 7.3 shows a set of spectra taken from one interferometer output port. For figure (a) white light illumination was used and the path difference was set to $s = 40\ \mu$m. According to equation 7.1, a sine-like modulation with a period $\Delta\lambda = 5.5$ nm is observed in the spectrum. When switching to discrete spectra the full power of this method becomes evident. In the top graph of figure 7.3(b) an unfiltered few-quantum dot spectrum is displayed with several well-separated spectral lines. It was obtained by blocking one interferometer arm. But when unblocking and exposing the lines to interference, it was possible to align the path difference for selectively switching on and off individual lines. This is the case in the lower four graphs, indicated by three arbitrary selected spectral lines. In the other interferometer output the opposite picture would be visible. In this way, an inventive spectral switch can also be achieved.

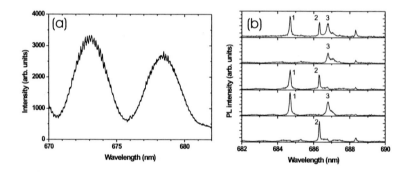

Figure 7.3.: *(a) Spectrum of a white light source observed through the Michelson interferometer. (b) Spectrum of a few InP quantum dots through the interferometer. In the top graph, one arm was blocked resulting in the original spectrum. In all five graphs, the intensity axes were equally scaled, for comparison. The numbers indicate three arbitrary chosen spectral lines, when visible, which depends on the path difference in the interferometer.*

7.1.2. Experimental demonstration using two-photon cascades

In chapter 6, multi-photon cascades in InP quantum dots were investigated, where it was found that they form a source of photon pairs, well separated in energy. Due to

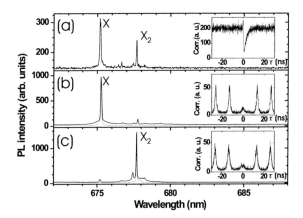

Figure 7.4.: *(a) Unfiltered spectrum of the quantum dot emission. X and X_2 denote the exciton and biexciton spectral line, respectively. The inset shows a continuous cross-correlation measurement between the two spectral lines. (b) Spectrum of the light emerging from one Michelson output port. (c) The same as in (b) but taken at the other output port. The gray curves have been obtained after filtering with a bandpass filter. In the insets, pulsed auto-correlation measurements of the filtered exciton and biexciton emission taken behind each output port are displayed.*

the strong correlation in their emission time, the time separation of the two photons lies within the transition lifetime. Figure 7.4(a) shows a spectrum of the emission from a single quantum dot. It was taken at one interferometer output but with one arm blocked. The emission from exciton (X) and biexciton (X_2) is clearly distinguished by a wavelength separation of $\Delta\lambda = 2.5$ nm. The assignment to the biexciton and exciton was done by looking on the excitation power dependency of the lines in combination with analysing cross-correlation measurements (inset of figure 7.4(a)). The excitation power of 30 pW/μm^2 was chosen to have similar line intensities.

By adjusting the arm length difference in the unblocked Michelson interferometer, a situation was achieved, where the exciton and biexciton lines show constructive interference in either output port and destructive in the other, respectively. The resulting spectra are displayed in figures 7.4(b) and (c) as black lines. In this way, the exciton and biexciton emissions were suppressed in the according port by 90.5% and 91.4%, respectively. The total transmission of each separated line was 77%, where the main losses result from reflections from the uncoated optical surfaces. For detecting single photon emission, 0.5-nm bandpass filters were placed behind each output to suppress background and excitation light, resulting in the gray spectra.

To compensate for mechanical drifts during long-timed correlation measurements, a control software was written to actively stabilize the Michelson interferometer on maximum count rates. The insets in figures 7.4(b) and (c) show the second-order coherence function obtained from the lines filtered that way. A clear suppression of the peaks around zero time delay is observed. The central-peak area is only 0.17 relative to the areas under the other peaks, which demonstrates the anti-bunched nature of the emission. This value is slightly larger than obtained in previous measurements (such as figure 4.3) due to the stronger excitation, necessary to also excite the biexciton. In this way, a single quantum emitter could have been used as a source for two distinct single photons.

In a further step, the setup was expanded by the part sketched in figure 7.2(c): The two output beams were, after filtering, coupled into two multi-mode fibres. A different length in the fibres provided a relative time delay of 6.6 ns (half the repetition time of the pulsed Ti:Sa laser). After outcoupling the two beams were merged and detected by the Hanbury Brown-Twiss detectors. In this way, the train of exciton photons was shifted in between the biexciton photon train. This enables the simultaneous observation of the two photon sources and leads to a stream of single photons with a doubled repetition rate. Ideally, a second, inversed Michelson arrangement will be used to merge the two beams leaving the fibres. However, for simplicity, a 50:50 beam splitter was used instead.

Figure 7.5(b) shows correlation measurements of the light, merged after coupling out of the fibres. The excitation power is set as in the measurements displayed in figure 7.4. For comparison, an exciton correlation function is again displayed in figure 7.5(a). Both figures exhibit the characteristics of a pulsed single-photon source. While the impinging photons in 7.5(a) have the familiar time separation of 13.2 ns, determined by the excitation laser repetition rate of 76 MHz, the photon stream in 7.5(b) possesses only half of the repetition time. Still, a clear anti-bunching is visible.

The last graph also demonstrates, how the maximum emission rate of a single-photon source based on spontaneous emission is limited. As the photodetectors cannot distinguish between the energy of the two photons, a similar correlation measurement would have been obtained if photons from only excitonic transitions had been recorded, but with a doubled excitation rate. In both situations the time period between excitation events approaches the spontaneous lifetime, which is on the order of 1 ns for the quantum dot transitions, as reflected by the peak widths. Thus, the peaks start to overlap and the photons cannot be assigned to the individual excitation pulses any more, which is vital for their use in quantum communication. The result in figure 7.5(b) is already on the onset of this process. However, in the presented kind of experiment, adjacent photons remain distinguishable with respect to their wavelength and an assignment of each photon to a pulse can be preserved.

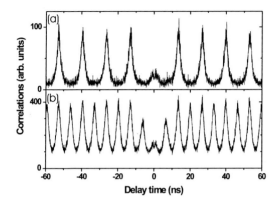

Figure 7.5.: *Intensity correlation (a) of the exciton spectral line and (b) of the multiplexed signal.*

Another interesting feature in figure 7.5(b) is the varying height of the peaks in the correlation measurement at higher delay times, which is especially pronounced at the reduced first peaks besides the time origin. Similar modulations of the peak height have also been observed in a conventional single photon source, where only one transition is used [SFV+04]. There, this effect has been attributed to different probabilities to populate the excitonic state from a charged or an uncharged quantum dot in case of off-resonant and resonant optical excitation, respectively. In the configuration presented here, the origin of correlations between adjacent peaks is more complex, since a start–stop event may be caused by either a biexcitonic photon followed by a delayed excitonic photon or vice versa. Correlations between the direct successive photons from a cascade have been discussed in the previous chapter. Additionally, capturing another charge carrier, as it was observed on this sample (see chapter 4.3), can change the charging state of the dot and influence correlations between neighbouring peaks as well as (on a smaller degree) between more distant peaks.

In order to study these peak height variations, Monte-Carlo simulations were performed, which are based on the cascade model sketched in figure 4.8. In contrast to the simulations of chapter 4.3, now, both exciton and biexciton decay times were recorded, with an additional 6.6-ns offset for the biexciton decay time to account for the fibre delay in the experiment. Figure 7.6 displays two typical results from this simulation. The decay rates that are used for the simulation are 0.71 GHz for the exciton and 0.83 GHz

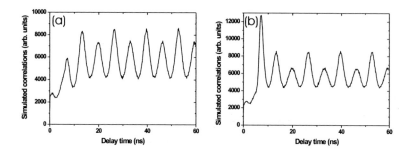

Figure 7.6.: *Results obtained by a Monte-Carlo simulation: (a) for exciting both charged and uncharged and (b) only uncharged exciton and biexciton states.*

for the biexciton decay, as extracted from the fits to the auto-correlation function of each transition (insets of figure 7.4). When exciting above the barrier band gap, as it is done in the experiments of this work, free charges are excited and trapped in the quantum dot, creating both charged and uncharged excitons and biexcitons. On the other hand, resonant excitation of higher quantum dot states, but below the continuum, will create mainly uncharged excitons and biexcitons. Figure 7.6 depicts results from numerical simulations for correlations between unequally (a) and equally (b) charged transitions, respectively. It can be seen that the numerical data reproduces well the anti-bunched first peak as well as the slightly alternating heights of the higher order peaks from the experimental data (figure 7.5(b)), whereas figure figure 7.6(b) displays features that have been observed similarly with quasi-resonantly excited InAs quantum dots [SFV+04].

7.2. Quantum key distribution

7.2.1. The BB84 quantum cryptography protocol

Cryptography is the science of encrypting and concealing information. Its main goal is the secure transmission of information to a receiver, while preventing a third party from reading it. Additional tasks also involve the authentication of the sender or of the message itself. First applications of cryptography are known since the early days of human history, with the use of unusual hieroglyphics or by rotating the letters in the alphabet (Caesar cipher). Modern techniques can be categorized in two groups: When using one-time-pads (or Vernam-ciphers), the encryption key, ideally made of a random character sequence, is at least as long as the message itself. It has to be previously distributed between the communication partners and is used only once. This system has been proofed secure, but facing the increasing information flood in today's communication, the necessary key exchange makes this system impractical for most applications. This problem is solved by asymmetric protocols, in which the message is encrypted by a publicly accessible key and decrypted by a private key, which only the legal receiver possesses. Here, the security arises from the assumed impossibility to crack the encryption algorithm within reasonable computing time. However, this security is illusive, as perpetually advanced computer power forces the use of steadily increasing key lengths.

In 1984, Ch. Bennett and G. Brassard discovered that, when encoding information to quantum states instead of classical bits, it is possible to securely exchange a random key over large distances, while revealing potential eavesdropping attacks [BB84]. In this way, they avoided both problems of classical cryptography: A one-time-pad for secure encryption can now be exchanged without the necessity of sender and receiver to meet before by still knowing that no third party possesses this key.

Figure 7.7 illustrates the principle of this so-called *BB84 protocol*. By convention, sender and receiver are referred to as *Alice* and *Bob*, respectively, while an eavesdropper is shortly called *Eve*. For the BB84 protocol two transmission channels are needed: one carrying quantum information from Alice to Bob (here, encoded in the polarization of single photons) and a publicly accessible classical channel, as for example a telephone line (figure 7.7(a)). In a first step, Alice sends Bob a stream of single photons, randomly polarized in horizontal ($|H\rangle$), vertical ($|V\rangle$), left-handed ($|L\rangle$) or right-handed ($|R\rangle$) circular direction (figure 7.7(b)), while Bob detects these photons also randomly in either the linear or circular basis (figure (c)).[1] Both keep track of the prepared and detected states, respectively (figure (d)). After the transmission of quantum information, Alice and Bob exchange publicly which bases they used for each qubit and select out only data with a common choice of bases, so that they agree on the prepared/measured state (figure (e)). After assigning these states to classical information, like $\{|H\rangle, |L\rangle\} \stackrel{\wedge}{=} 0$ and $\{|V\rangle, |R\rangle\} \stackrel{\wedge}{=} 1$, they end up with a common set of random numbers, figure 7.7(f). After excluding an eavesdropping attack (see below), these random numbers can be unconcernedly used as a secure one-time-pad for classical message encryption.

The fascinating part of the protocol is to see how eavesdropping can be monitored. Wiretapping the classical channel alone does not give the necessary information to crack the encrypted message, as here, only the polarization bases, but no information about the prepared state itself, is exchanged. Thus, Eve cannot help but to also measure the polarization in the quantum channel by detecting the photons. At this point the importance of using single-photon states becomes evident. As Eve's detection process ultimately annihilates this sole photon, she has to cover her existence by faking the

---

[1] Alternatively any other combination of two orthogonal bases, such as horizontal/vertical and ±45°-polarization, can be used.

photon state. But due to quantum mechanics' No-Cloning theorem [WZ82], she cannot copy the original state with 100% fidelity and must make a minimum amount of mistakes in preparing the faked state. The resulting increase of transmission errors can be detected by Alice and Bob, before sending the real message, by comparing a part of their keys, first. Thus, the strength of quantum cryptography lies in the ability to discover eavesdropping attacks and to abort communication before private data is transmitted.

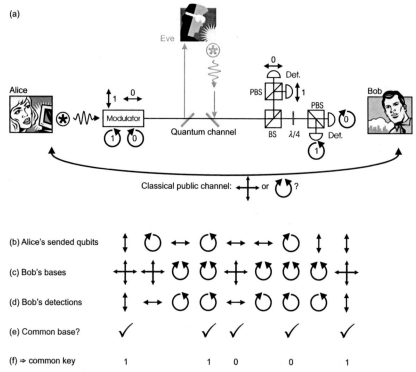

Figure 7.7.: *(a) Illustration of the components for BB84 quantum key distribution. The single photon sources are symbolized by encircled asterisks. As a modulator, an electro-optical modulator to vary the polarization can be used. The beam splitter (BS) randomly distributes the photons to the detections of the two bases. A quarter-wave plate (λ/4) converts circular to linear polarized light and the polarizing beam splitters (PBS) separate the two components of a polarization basis and lead the photons to the photodetectors (Det.). (b)–(f) is an example transmission, as explained in the text.*

Until today, several further protocols have been suggested, among them for instance the Ekert protocol using entangled photon pairs [Eke91]. A review and detailed discussion is given by Gisin et al. [GRT+02]. Long distance experiments have been successfully realized with weak coherent laser pulses [KZH+02, GRT+02] and down-converted entangled photon pairs [JSW+00]. Realizations of the BB84 protocol with single-photon states were performed using diamond defect centres [BBG+02] and single quantum dots [WIS+02].

7.2.2. Multiplexing the BB84 protocol – discussion and demonstration

It can be seen, that in the BB84 protocol, information is stored solely in the photon's polarization whereas the exact wavelength is unimportant. Thus, multiplexing as described previously can provide an increased communication bandwidth without loss of security. Figure 7.8(a) shows a possible implementation of interferometric multiplexing into the BB84 protocol. On Alice's side, a cascaded photon source, such as a single quantum dot, provides two closely emitted single-photons with different energies upon each excitation pulse. In a first Michelson interferometer, the two photons will be separated. In the same way as in the conventional protocol, polarizers define a fixed polarization[2] and electro-optic modulators (EOMs) randomly modulate between (H, V, L, R) polarization for each photon. At the latest here, any polarization correlation between the two photons is destroyed, thus providing independent qubits. An inversely aligned Michelson interferometer recombines the two photons in a single channel for transmission. On the other side Bob uses the same arrangement to separate the photons. A second set of EOMs is used to randomly change the bases. Polarizing beam splitters in combination with two APDs detect the polarization state of each photon. In this way, the transmission rate will be doubled compared to a protocol using only one photon per pulse.

To demonstrate this application, a simplified proof-of-principle experiment, as sketched in figure 7.8(b) was set up: The excitonic and biexcitonic photons from the quantum-dot single-photon source were separated with a Michelson interferometer, fibre-coupled and delayed, as done for the experiments in section 7.1. As a simplification the second add/drop Michelson (which combines the photons again) was replaced by a 50:50-beam splitter. With this, the photon pulse rate was doubled, but the average photon number per pulse was halved. Deviating from the proposed scheme, Bob's detection consisted of a second EOM, an analysing polarizer and an APD, with the EOM randomly switched between the two bases. Since only in one fourth of the cases Bob measures in the same state as Alice, this results in a reduction of the effective count rate of 50% compared to a scheme with two detectors. In this configuration, two-photon events can create a possible insecurity, but in our setup the collection efficiency is estimated to be $p \approx 10^{-3}$ (see

---

[2]For already polarized photons, the polarizers would be ideally aligned for optimum transmission.

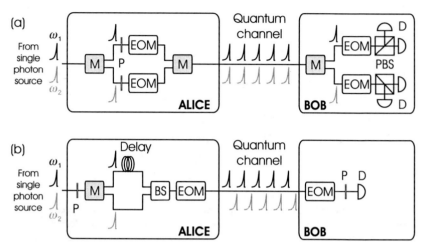

Figure 7.8.: *Possible implementations of the multiplexer into the BB84 protocol. In both schemes the polarization is modulated between straight and circular polarization using the polarizer P with an electro-optical modulator (EOM). Bob's detection side is realized by another EOM, a polarizing beam splitter PBS and detectors D. ω_1 and ω_2 indicate the two energies of the photons and M is the Michelson arrangement. In (a), the two photons are recombined without time delay, in (b) a time delay is introduced and the two beams are then recombined with a beam splitter BS, like it is done in the experiment.*

section 3.6). The probability to collect two adjacent photons is $p_{11} = p^2 \approx 10^{-6}$ and thus much smaller than the probability to collect one and loose the next, $p_{10} = p(1-p) \approx 10^{-3}$. At higher collection efficiency, the setup in figure 7.8(a) is favoured.

The transmission distance was 1 m. For exciting the quantum dots, a pulsed diode laser ($\lambda = 635$ nm, pulse width 125 ps) with a repetition rate of 10 kHz was used, which was adopted to the modulation rate of the EOM drivers. These drivers consisted of a digital-to-analogue converter, steered by a computer card, and a subsequent high-voltage amplifier to supply the EOMs with the half- and quarter-wave voltages. A rectangular voltage signal acts as a trigger for the laser pulses, the EOM switching and the detection gate for acquiring Bob's detection events. The trigger and the detection gate were shifted towards the end of the EOM switching period, in order to avoid initial voltage spikes by the EOM driver to affect the photon polarization. The presence of these spikes dictated the maximum modulation rate. The choice of the random bases and data acquisition were controlled by a Labview programme. An improved software-based random number

Figure 7.9.: *Visualization of the quantum key distribution. After exchanging the key, Alice encrypts image (a), a photography of Berlin's skyline taken from Hausvogteiplatz, and sends the encrypted image (b) to Bob. After decrypting it with his key, he obtains image (c).*

generator provided the randomness of the bases.

Due to the mentioned simplifications, the achieved transmission rates are smaller than expected in the optimized configuration from figure 7.8(a). Nevertheless, it successfully demonstrates a doubling of the transmission rate with respect to the same setup without multiplexing.

In the images of figure 7.9, the results of a quantum key distribution are visualized. In a first step Alice and Bob exchanged quantum information resulting in a common sequence of random bits. A series of random number tests checked and confirmed the randomness of the key. This key then encrypted image 7.9(a) by applying an exclusive-OR (XOR) operation between every bit of image and key. This results in image 7.9(b) in which the randomness of the key was transferred. Then, image 7.9(b) was submitted classically to Bob who decrypted it by applying another XOR operation with his received key, yielding image 7.9(c).

Altogether the experiment was run with the following parameters: After the electronic gating of Bob's detector signals, the rate of usefully exchanged photons is found to be 30 s^{-1} whereas the dark count rate is reduced to 0.75 s^{-1}. The probability to transmit photons through the two EOMs with crossed polarizations was measured to be 6.8%. After comparing Alice's and Bob's keys an error rate of 5.5% was found. The presence of transmission errors leads to the necessity of error correction. This requires the exchange of redundant data which opens an eavesdropping loophole for gaining partial information of the message. With the experimental parameters, the number of secure bits per pulse following Lütkenhaus [Lüt00] is 5×10^{-4}, which is a typical value for current single-photon quantum cryptography experiments ($\approx 1 \times 10^{-3}$ secure bits per pulse, see references [BBG$^+$02, WIS$^+$02]).

The Michelson add/drop filter might also find applications in linear optical quantum computation (LOQC) [KLM01]. Since gates in LOQC have only limited success probabilities, parallel processing may increase the efficiencies of the gate or at least improves the statistical significance of a computational result. As pointed out in chapter 6.3, multi-photon decay cascades in single quantum dots have also been proposed as a possible source for entangled photon pairs [BSP+00]. Here, the photons are polarization entangled and in general not energetically degenerated. The method which was demonstrated here allows the spatial separation of the two photons without destroying their entanglement. Thus they can be subsequently used in a multitude of experiments and applications.

8. Outlook

Single-photon quantum key distribution is only one possible application of single quantum emitters and of non-classical light. Other quantum cryptography protocols, but also quantum algorithms require entangled photon pairs or photons in a lifetime-limited mode. Coupling of quantum dots to microcavities might give important advantages and opens the way to new regimes of light–matter interaction [RSL+04, YSH+04]. Other applications of single-photon sources might be quantum memories using electromagnetically induced transparency (EIT) [FYL00], which requires a very narrow spectral photon bandwidth. This chapter displays future perspectives of this work. First results of a novel approach for controlled positioning of quantum dots to microsphere cavities are presented. Possible developments of the project towards indistinguishable photons and entangled-photon pairs are discussed.

8.1. Single quantum dots on micro-tips – towards coupling of a single quantum dot to a microsphere cavity

One attempt to optimize the control over the emitted modes of single photons is to couple the quantum emitter to microcavities, which enhances the emission into cavity modes and additionally reduces the radiative lifetime [GSG+98, PSV+02].

For optimum coupling, however, very precise positioning of the quantum emitter to the anti-nodes of the spatial cavity mode distribution is necessary. This seems to be in contradiction to the self-organized growth process of quantum dots.

Here, first results of an approach are described, where we followed the strategy of prestructuring the substrate [CWL+04] and extended it to the growth on sharp tips [ZAH+04]. For this, tips were first etched into the substrate and quantum dots were fabricated by epitaxial growth. This approach has several advantages. First, it avoids any post-growth processing and can therefore be expected to result in higher quality samples, as no defects are created through the processing of the sample after the growth. Second, quantum dots on tips are spatially isolated and can be manipulated, for example optically excited, without the need of high spatial resolution. Finally, a tip structure with a single quantum dot represents an active probe similar to a functionalized AFM

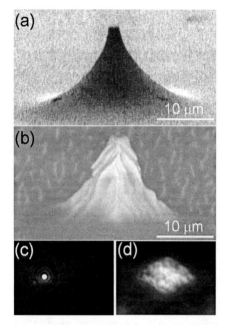

Figure 8.1.: *SEM image of a GaAs tip before (a) and after (b) overgrowth. (c) Micro-photoluminescence image of a single quantum dot on a tip taken through a narrow band-pass filter. (d) White light microscope image of a tip from the top. The widths of the two bottom images are 22 μm each.*

(Atomic Force Microscopy) [NVL97] or SNOM (Scanning Near-field Optical Microscopy) [MHM⁺00] cantilever. An implementation of such a tip structure in a scanning apparatus would be straightforward [GBS01]. Such a scanning probe would be very attractive both for high resolution microscopy as well as for controlled experiments at the level of a single quantum emitter.

In our structures we started with SiO disks with 200 nm thickness, diameters of 10 μm and spacings of 150 μm which were fabricated by photolithography and buffered HF etching on undoped GaAs (001) substrates. Next, the samples were wet etched in a solution of H_2SO_4:H_2O_2:H_2O (in proportion 8:1:1) to form sharp tips with sizes of a few micrometres; the remaining SiO masks were finally removed with a buffered HF solution. The GaAs etching is slightly anisotropic and results in square shaped tips. These tips were about 2 μm on a side and 6 μm tall. Following a 10 s etch in

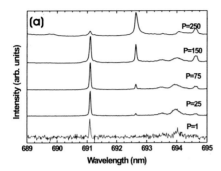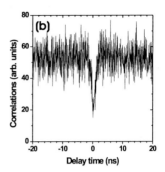

Figure 8.2.: *(a) Spectra of a single dot on a tip edge at different excitation intensities. The spectra were normalized and offsets were added for clarity. The excitation power unit is $P = 0.4$ nW/μm^2. (b) Photon correlation measurements of the exciton emission line from the same dot.*

HF, the samples were loaded into the ultra-high vacuum epitaxy system for quantum dot growth.[1] The quantum dots were grown using a gas-source molecular beam epitaxy (GSMBE).[2] After oxide desorption a 100 nm GaAs buffer layer was grown, followed by 200 nm of $Ga_{0.48}In_{0.52}P$. Subsequently, 1.8 monolayers of InP were deposited, to form the quantum dots and the resulting structures were capped with 50 nm $Ga_{0.48}In_{0.52}P$.

Figure 8.1 shows a Scanning Electron Microscope (SEM) image of a tip before (a) and after (b) the overgrowth. As can be seen, the overgrowth does not result in a smooth and flat surface. The thicknesses of buffer, quantum dot and capping layers given above are thus nominal values which refer to growth on a flat surface. Figure 8.1(c) shows a micro-photoluminescence image of a single quantum dot on a tip taken through a narrow bandpass filter centred at 690 nm. Individual dots are easily resolved predominantly at the edges of the tips. Also, a very low quantum dot density on the slopes of the tips was observed, while the density of dots in the planar regions between the tips is high. It was observed that dots consistently nucleate at the edges of the tip. The shape of a tip is revealed in figure 8.1(d) where the tip shown in figure (c) is imaged with white light.

The optical performance of the emission of a quantum dot on a tip is shown in figure 8.2. In graph (a) the power dependent spectrum, showing an exciton and a biexciton

---

[1] We thank E. Wiebicke from Paul Drude Institut, Berlin, for assistance with the tip etching.
[2] The growth of the quantum dots was performed by F. Hatami in the group of Prof. W. T. Masselink from Humboldt-Universität zu Berlin.

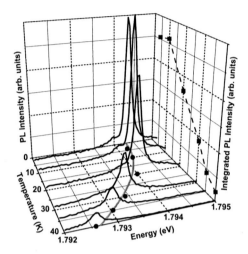

Figure 8.3.: *Single dot exciton spectra taken at different temperatures. The black circles indicate the peaks positions and the black squares show the integrated intensity. The gray line is the expected temperature dependence and the dashed line shows the evolution of the integrated emission intensity with increasing temperature.*

spectral line at different excitation powers, is displayed. Note that another advantage of having single dots on sharp tips is, that only the dot under investigation is in the focal plane of the collection optics, yielding a better signal-to-background ratio. Figure 8.2(b) shows the second-order coherence function of the exciton line. The deep anti-bunching dip demonstrates single-photon emission, with time scales comparable to InP quantum dots on a planar substrate (see chapter 4.1). These experiments were performed at 5 K.

In applications, such as single quantum dot lasers or LEDs with modified spontaneous emission, it is required to couple the light of a single emitter to an optical cavity with a narrow resonance. Thus, spectral tuning is essential for a single quantum dot on a tip. In this perspective, temperature dependent measurements on the exciton emission from a single dot were performed from 5 K up to 40 K (figure 8.3). The exciton emission was studied under constant excitation intensity and all spectra were taken with the same integration time (1 s). It was observed that a tuning range of about 250 μeV (60 GHz) can be achieved by increasing the temperature from 5 K up to 20 K. At higher temperatures, the emission intensity decreases drastically but emission is still observed at 40 K with a shift of 1.65 meV (400 GHz). The emission linewidth increases from

150 μeV (resolution limited) at 5 K to 250 μeV at 30 K. The vertical positions of the black squares are proportional to the integrated intensity of the exciton emission. The temperature shift is accounted for by the temperature dependence of the bandgap. In figure 8.3, the black dots represent the peak position as a function of temperature and the gray line is the expected temperature dependence of the bandgap obtained with a fit of the form $E(T) = E_0 - \alpha T^2/(T+\beta)$, with $\alpha = 4.91 \times 10^{-4}$ eV/K and $\beta = 400$ K, which is in good agreement with previous temperature studies performed on single quantum dots made of the same material system [MT02].

This configuration now enables precise positioning of a single quantum dot to a microcavity. For this, cavities based on glass microspheres are very interesting systems, as they combine today's highest achievable quality factors with small mode volumes.[3] In such a system, the cavity is formed by whispering-gallery modes (WGMs) generated by total internal reflection. In such modes, part of the WGMs leak out of the sphere as an evanescent field, which enables coupling of external emitters to the cavity. Successful coupling of nanocrystals and dye-doped beads was reported in reference [Göt03]. Such microspheres are produced by melting a tapered glass-fibre in the focus of a CO_2 laser, whose emission wavelength ($\lambda = 10.6$ μm) is strongly absorbed by glass. With this procedure, silica microspheres can be routinely produced, having diameters between 20–300 μm, which are adjustable during the melting process. An image of such a sphere is shown in figure 8.4(a)

The difficulty in coupling microspheres to quantum dots on tips is to compose an experimental setup, in which the quantum dots remain in a cryogenic environment, while the microsphere can be fine positioned above the quantum dot and while a confocal microscope enables the efficient detection of the quantum dot photoluminescence. For first observations, the setup sketched in figure 8.4(b) was chosen. In this setup, the top cover of the continuous-flow cryostat is replaced by a self-made head containing a metal spring screwed on the top plate. With a piezoelectric translator pressing on the spring, it is adjustable from the outside in the vertical direction. This enables the vertical alignment of the sphere relative to the tip, while horizontal positioning is performed by the cryostat step motors. External translation stages place the whole arrangement below the microscope. While controlled under a microscope, the sphere can be glued on the spring so that the plane of the WGMs is vertically oriented. In this way it becomes possible to position a quantum dot on a tip near a glass microsphere with a precision determined by the cryostat step motors (0.1 μm).

Using this setup, first evidence of coupling of the emission of quantum dots to a microsphere cavity could be observed. Figure 8.4(c) shows the spectrum of the stray light collected tangentially from a sphere after placing it on an overgrown tip. Due to the weak signal, a comparably high excitation intensity was needed, resulting in a strongly broadened spectrum. On top of the spectral envelope, a periodic double peaked structure

---

[3]Quality factors up to 10^9 and mode volumes of 250–8000 μm^3 are typical values [Göt03].

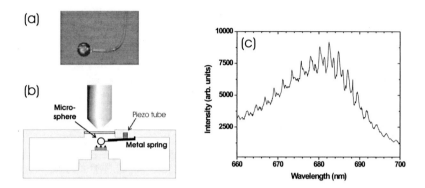

Figure 8.4.: *(a) Microscope image of a glass microsphere with a diameter of 60 μm (from [Göt03]), (b) Setup of the modified cryostat, (c) spectrum of the stray light of a 45 μm microsphere placed on top of a tip containing several InP quantum dots. The offset was adjusted, so that the wavelength axis corresponds to the electric noise level of the spectrograph CCD camera.*

can clearly be observed, where the peaks can be identified as transversal electric (TE) and transversal magnetic (TM) modes of the cavity [Göt03]. The spectral periodicity of 2.3 nm is in agreement with the free spectral range (FSR) of 2.24 nm, expected from the used sphere with a diameter of 45 μm. When lowering the excitation intensity, the broad envelope of the spectrum transforms to a few sharp but very weak spectral lines. However, in this weak-excitation regime no mode structure was observed any more.

8.2. Indistinguishable photons

The generation of lifetime-limited or indistinguishable photons is vital for the implementations of purly optical quantum gates, as these gates are based on two-photon second-order interference effects. This effect, first measured by Hong et al. [HOM87], occurs when two identical photons enter the two sides of a 50:50 beam splitter at the same time. In this special case, quantum mechanics predicts, that the two photons will leave one of the two beam splitter outputs with equal probability *in pairs*. A correlation measurement between the outputs will result in a drop of coincidences, called the Hong-Ou-Mandel dip. In order to observe this dip, the measurement of lifetime-limited spectra is a prerequisite, although it is not a strict condition, as for example Santori et al. [SFV+02] reported a Hong-Ou-Mandel dip using a single photon source with a clearly broadened spectrum.

Still, in order to succeed in two-photon interferences using quantum dot emitters, a spectral bandwidth at least close to the homogeneous linewidth is required. There are several strategies that can be followed. Resonant population of excited exciton states strongly reduces the generation of free charge carriers in the sample wetting layer and with this suppresses effects that arise from their interaction with the quantum excitons. It has also been shown [KVC+02] that the photon linewidth was reduced when the quantum dot was excited close to resonance, as the energy deposited in the system is completely converted into light, leaving no excess energy to excite phonons. For the InP quantum dot system, this is a critical task as the excited (p-shell) exciton states are expected to be about 50 meV below the exciton ground state wavelength, which is already close to the transitions of the wetting layer. Thus, a very careful tuning of the laser line to the excited state is required. An alternative method is the resonant two-photon excitation of the biexciton state [FBA+04]. This is especially interesting in II-VI type quantum dots, which exhibit a large biexciton binding energy (around 20 meV), permitting efficient separation of photoluminescence from excitation stray light.

Lifetime limited linewidths are easier obtained, if the lifetime is already short, so that dephasing events have no time to occur during the emission of the photon. II-VI quantum dots already have a short lifetime (around 200 ps), compared to III-V quantum dots (around 1 ns). Quantum dots caused by interface defects in thin quantum wells have shown to possess very large oscillator strength and therefore a very short lifetime (around 20 ps, [HVG+03]).

Further improvements can be obtained by coupling a quantum dot to a microcavity, where the increased mode density enhances the spontaneous emission rate in the cavity mode, thus reducing the lifetime. This effect is characterized by the so-called Purcell factor [Pur46], describing the ratio between the emission rate of a cavity-coupled emitter to the rate of an emitter in free space. It is given by $\gamma/\gamma_0 \propto Q\lambda^3/V_0$, where Q is the quality factor and V_0 the mode volume of the cavity mode. Using micropillar cavities, a quantum dot lifetime reduction by factors larger than 5 has been reported [GSG+98, PSV+02]. The extraordinary high quality factors of microsphere resonators enable Purcell factors of several hundreds with typical parameters (sphere diameter: 30 μm, $Q \approx 10^6$, $\lambda \approx 700$ nm)

8.3. Entangled photon-pairs

The generation of entangled photon-pairs is a fascinating application of single quantum emitters. In a proposal of Benson et al. [BSP+00] this is achieved by the polarization dependent decay of a biexciton–exciton cascade over two indistinguishable decay channels, as described in chapter 6.3.

Two main obstacles have prevented its experimental realization so far: First, during the decay cascade, there emerges a sensitive entanglement between the first emitted photon and the remaining exciton, which is very susceptible to decoherence. In the same way as in the previous topic, resonant excitation and lifetime reduction lead to improved coherence properties. Second, in general the degeneracy of the two involved exciton states of different spin is lifted, which principally allows to distinguish the two decay channels and destroys entanglement. Applying electric fields results in a spin dependent Stark shift, making it possible to compensate the exciton level splitting. In this regard, charged quantum dots are especially exciting: in this system, the Kramers theorem predicts that in absence of external fields, states with a half-integer spin, such as a negatively charged ground-state exciton $(S_{\text{tot},z} \in \{\pm 3/2, \pm 1/2\})$, are twofold degenerated. This intrinsic degeneracy makes charged quantum dots promising for the generation of entangled-photons. Care has to be taken, as the decay of the charged biexciton may result in an excited trion (charged exciton) state that relaxes nonradiatively to the trion ground state, which is a process where decoherence may occur. This could be avoided by observing the direct radiative decay of the excited trion [UBM$^+$04].

In another type of photon entanglement, the two photons of a cascade are in a coherent superposition of being emitted at two different times, resulting in a state of the form $|\text{early}\rangle_{\omega_1}|\text{early}\rangle_{\omega_2} + \exp(i\phi)|\text{late}\rangle_{\omega_1}|\text{late}\rangle_{\omega_2}$. This so-called time-bin entanglement was first developed for down-conversion photon-pairs [BGT$^+$99] and recently proposed for quantum dot cascades [SP04]. In this scheme, the biexciton state is populated by two delayed pump pulses. The two pump pulses must have a fixed relative phase, which can be achieved in a Mach-Zehnder interferometer, and need carefully chosen intensities to realize a maximum entangled state. This creation of the succeeding photon pair by biexciton decay has to happen coherently, such that no information about the emission time of the photon pair is stored anywhere in the emitting system or in the environment. Under this condition, the generated state is a superposition of both photons being emitted at the earlier or later time. Time-bin entanglement is particularly well suited for long-distance transmission in optical fibres as it is insensitive to polarization fluctuations.

Appendix A.

Solution of the Rate Model

Here, the solution of the rate model, equation (6.4), used in chapter 6.2.1 and involving the biexciton–exciton cascade, is derived. This model is based on the following set of rate equations for the population probability densities:

$$\frac{d}{dt}\mathbf{n}(t) = \begin{pmatrix} -\tau_E^{-1} & \tau_1^{-1} & 0 \\ \tau_E^{-1} & -(\tau_E^{-1} + \tau_1^{-1}) & \tau_2^{-1} \\ 0 & \tau_E^{-1} & -\tau_2^{-1} \end{pmatrix} \mathbf{n}(t) = \hat{N}\mathbf{n}(t),$$

(A.1)

$$\text{with} \qquad \mathbf{n}(t) = (n_G(t), n_1(t), n_2(t)).$$

This model differs from that for equation 6.2 in a way that its highest state is given by the biexciton. τ_1 and τ_2 are the exciton and biexiton lifetimes, respectively and τ_E^{-1} denotes the excitation rate. $n_G(t)$, $n_1(t)$ and $n_2(t)$ are the population densities for the ground, exciton and biexciton state, respectively.

The eigen-values of the rate matrix $\hat{N}$ are:

$$\lambda_0 = 0,$$
$$\lambda_{1,2} = -\frac{1}{2}\tau_E^{-1}\left(\eta_1 + \eta_2 + 2 \pm \sqrt{(\eta_1 - \eta_2)^2 + 4\eta_1}\right),$$

(A.2)

with the abbreviations $\eta_{1,2} = \tau_E/\tau_{1,2}$. The eigen-vectors are:

$$\mathbf{l}_0 = (\eta_1\eta_2, \eta_2, 1),$$
$$\mathbf{l}_{1,2} = (-\eta_2 - 1 - \lambda_{1,2}\tau_E, \eta_2 + \lambda_{1,2}\tau_E, 1).$$

(A.3)

In the diagonal base, (A.1) looks as: $d\mathbf{m}(t)/dt = \hat{M}\mathbf{m}(t)$ with $\hat{M} = \mathbf{1}(\lambda_0, \lambda_1, \lambda_2)$, where $\mathbf{1}$ is the unit matrix. The solution of this is $m_i = \text{const.} \times \exp(\lambda_i t)$, $i \in \{G, 1, 2\}$. With the unitary transformation matrix $\hat{U} = (\mathbf{l}_0, \mathbf{l}_1, \mathbf{l}_2)$ we get $\hat{N} = \hat{U}^{-1}\hat{M}\hat{U}$ and $\mathbf{n}(t) = \hat{U}\mathbf{m}(t)$. Finally, we obtain:

$$
\begin{aligned}
n_G(t) &= A\eta_1\eta_2 + B(-\eta_2 - 1 - \lambda_1\tau_E)e^{\lambda_1 t} + C(-\eta_2 - 1 - \lambda_2\tau_E)e^{\lambda_2 t}, \\
n_1(t) &= A\eta_2 + B(\eta_2 + \lambda_1\tau_E)e^{\lambda_1 t} + C(\eta_2 + \lambda_2\tau_E)e^{\lambda_2 t} \quad \text{and} \\
n_2(t) &= A + Be^{\lambda_1 t} + Ce^{\lambda_2 t}.
\end{aligned}
\tag{A.4}
$$

Here, B and C are integration constants, while $A = 1/(1 + \eta_2 + \eta_1\eta_2)$ gives the steady state populations and is determined by the condition $n_G(t)+n_1(t)+n_2(t) \equiv 1$ for all t. As the eigen-values are negative, $\lambda_{1,2} < 0$, the solutions are a sum of decaying exponentials. This is also true for the more extended rate model from equation 6.2.

The shape of the auto- and cross-correlation function is given by the investigated transitions. The transition observed by the start detector of the Hanbury Brown-Twiss system defines the initial conditions, while the count rate of the stop transition is proportional to the measured population density.

A.1. Exciton auto-correlation function

For the auto-correlation functions, both start and stop detector observe the same transition, in this case the exciton decay. Its detection projects the quantum dot in the ground state. Thus, the initial conditions are $n_G(0) = 1$ and $n_1(0) = n_2(0) = 0$. Using this in the solutions (equation (A.4)) specifies the integration constants:

$$
\begin{aligned}
B_1 &= \frac{\lambda_2}{\lambda_1 - \lambda_2}A \quad \text{and} \\
C_1 &= \frac{\lambda_1}{\lambda_2 - \lambda_1}A.
\end{aligned}
\tag{A.5}
$$

As the second-order coherence function is proportional to the probability that a photon is detected at time t after a first at time 0 was observed, which is again proportional to the exciton population density under the above initial conditions. Thus,

$$
g_{11}^{(2)}(t) \propto n_1(t) \propto N\left(1 + \frac{B_1}{A}(1 + \lambda_1\tau_2)e^{\lambda_1 t} + \frac{C_1}{A}(1 + \lambda_2\tau_2)e^{\lambda_2 t}\right).
\tag{A.6}
$$

The indices of the correlation function $g_{ij}^{(2)}$ indicate the observed start and stop transition (which are identical for auto-correlations). N is a proportional constant. When looking at the population $n_1(t)$, it is $N[n_1] = A\eta_2$ and for $g_{11}^{(2)}(t)$ one has to set $N[g_{11}^{(2)}] = 1$. For fitting to experimental data, it is just an open parameter.

A.2. Biexciton auto-correlation function

Here, the initial conditions are $n_1(0) = 1$ and $n_G(0) = n_2(0) = 0$, thus:

$$B_2 = \frac{1 + \eta_2 + \eta_1\eta_2 + \lambda_2\tau_E}{(\lambda_1 - \lambda_2)\tau_E} A,$$

$$C_2 = \frac{1 + \eta_2 + \eta_1\eta_2 + \lambda_1\tau_E}{(\lambda_2 - \lambda_1)\tau_E} A \tag{A.7}$$

and

$$g_{22}^{(2)}(t) \propto n_2(t) \propto N\left(1 + \frac{B_2}{A}e^{\lambda_1 t} + \frac{C_2}{A}e^{\lambda_2 t}\right), \tag{A.8}$$

with $N[n_2] = A$ and $N[g_{22}^{(2)}] = 1$.

A.3. Exciton–biexciton cross-correlation

The cross-correlation function has to be separated into two parts: For positive times, for example, the detection from the exciton transition starts the measurement and the biexciton decay stops it, whereas for negative times, start and stop are inverted. Thus, both cases apply to different initial conditions as well as to different observed population densities.

Bunching side (start: biexciton, stop: exciton)

As the biexciton starts the measurement and prepares the dot in the exciton state, the initial conditions are the same as for the biexciton auto-correlation, $n_1(0) = 1$ and $n_G(0) = n_2(0) = 0$, leading to the coefficients (A.7), whereas the stopping exciton decay reads out the exciton population:

$$g_{21}^{(2)} \propto n_1(t) \propto N\left(1 + \frac{B_2}{A}(1 + \lambda_1\tau_2)e^{\lambda_1 t} + \frac{C_2}{A}(1 + \lambda_2\tau_2)e^{\lambda_2 t}\right). \tag{A.9}$$

The normalization is $N[n_1] = A\eta_2$ and $N[g_{21}^{(2)}] = 1$.

Anti-bunching side (start: exciton, stop: biexiton)

Now, the exciton decay starts the measurement, leaving behind an empty quantum dot, $n_G(0) = 1$ and $n_1(0) = n_2(0) = 0$ with initial conditions (A.5), while the biexciton decay stops it and detects the biexciton population:

$$g_{12}^{(2)}(t) \propto n_2(t) \propto N\left(1 + \frac{B_1}{A}e^{\lambda_1 t} + \frac{C_1}{A}e^{\lambda_2 t}\right), \tag{A.10}$$

with $N[n_1] = A$ and $N[g_{12}^{(2)}] = 1$.

Notice, that in all discussed cases the factors in front of the exponentials are only dependent on the ratio $\eta_{1,2}$ between excitation and transition lifetimes.

A.4. Accounting for detector resolution

Due to the resolution w of the Hanbury Brown-Twiss setup, measured structures of the correlation functions at time scales smaller than $\sim w$ are washed out. To take this into account for the calculated functions, the model was convoluted with a Gaussian distribution of width w, which will be described now.

In the experiment, a change of start and stop detector leads to a change of the sign of time t, so that the total correlation function is set together by the positive and negative side using step functions $\theta(t)$ (remember that $\lambda_i < 0$):

$$
\begin{aligned}
g_{\text{tot}}^{(2)}(t) &= g_{\text{neg}}^{(2)}(-t)\theta(-t) + g_{\text{pos}}^{(2)}(t)\theta(t) \\
&= 1 + \sum_i \alpha_i e^{-\lambda_i t}\theta(-t) + \sum_i \beta_i e^{+\lambda_i t}\theta(t).
\end{aligned} \tag{A.11}
$$

The subsequent convolution of the exponentials with a Gaussian yields:

$$
\begin{aligned}
E(t,\lambda,w) &= e^{\lambda t}\theta(t) \otimes G_w(t) = \int_0^\infty e^{\lambda t'} G_w(t - t')\, dt' \\
&= \frac{1}{2} e^{\lambda t + \lambda^2 w^2/2} \left(1 + \text{erf}\left(\frac{t + w^2\lambda}{\sqrt{2}w}\right) \right),
\end{aligned} \tag{A.12}
$$

Here, $G_w(t) = (\sqrt{2\pi}w)^{-1} \exp(-x^2/(2w^2))$ is the Gaussian distribution of width w and $\text{erf}(t)$ is the error function. It can be seen, that these functions are not simply cut off at the time origin, but reach into the negative side. The washed out correlation function now reads:

$$
g_{\text{tot},w}^{(2)}(t) = g_{\text{tot}}^{(2)}(t) \otimes G_w(t) = 1 + \sum_i \alpha_i E(-t,\lambda_i,w) + \sum_i \beta_i E(+t,\lambda_i,w). \tag{A.13}
$$

Finally, the only thing left to do is to choose the coefficients α_i and β_i according to the configuration of start and stop detector.

Bibliography

[AGR82]   Alain Aspect, Philippe Grangier, and Gérard Roger, *Experimental realization of Einstein-Podolsky-Rosen-Bohm Gedankenexperiment: A new violation of Bell's inequalites*, Phys. Rev. Lett. **49** (1982), 91–94.

[AHF+02]   I. Akimiov, A. Hundt, T. Flissikowski, and F. Henneberger, *Fine structure of the trion triplet state in a single self-assembled semiconductor quantum dot*, Appl. Phys. Lett. **81** (2002), 4730–4732.

[AHS+04]   Thomas Aichele, Ulrike Herzog, Matthias Scholz, and Oliver Benson, *Single-photon generation and simultaneous observation of wave and particle properties*, Proc. AIP on Foundations of Probability and Physics **750** (2004), 35–41, see also arXiv:quant-ph/0410112.

[ARB04]   Thomas Aichele, Gaël Reinaudi, and Oliver Benson, *Separating cascaded photons from a single quantum dot: Demonstration of multiplexed quantum cryptography*, Phys. Rev. B **70** (2004), 235329(1–5).

[AWJ+00]   M. V. Artemyev, U. Woggon, H. Jaschinski, L. I. Gurinovich, and S. V. Gaponenko, *Spectroscopic study of electronic states in an ensemble of close-packed CdSe nanocrystals*, J. Phys. Chem. B **104** (2000), 11617–11621.

[AZB+03]   Thomas Aichele, Valéry Zwiller, Oliver Benson, Ilya Akimov, and Fritz Henneberger, *Single CdSe quantum dots for high-bandwidth single-photon generation*, J. Opt. Soc. Am. B **20** (2003), 2189–2192.

[AZB04]   Thomas Aichele, Valéry Zwiller, and Oliver Benson, *Visible single-photon generation from semiconductor quantum dots*, New J. Phys. **6** (2004), 1–13.

[BB84]   Charles M. Bennett and Gilles Brassard, Proc. IEEE Int. Conference on Computers, Systems and Signal Processing in Bangalore, India, IEEE, New York, 1984, p. 175.

[BBG+02]   Alexios Beveratos, Rosa Brouri, Thierry Gacoin, André Villing, Jean-Philippe Poizat, and Philippe Grangier, *Single photon quantum cryptography*, Phys. Rev. Lett. **89** (2002), 187901(1–4).

[BDC+98] H.-J. Briegel, W. Dür, J. I. Cirac, and P. Zoller, *Quantum repeaters: The role of imperfect local operations in quantum communication*, Phys. Rev. Lett. **81** (1998), 5932–5935.

[BEG+98] N. H. Bonadeo, J. Erland, D. Gammon, D. Park, D. S. Katzer, and D. G. Steel, *Coherent optical control of the quantum state of a single quantum dot*, Science **282** (1998), 1473–1476.

[BF02] M. Bayer and A. Forchel, *Temperature dependence of the exciton homogeneous linewidth in $In_{0.60}Ga_{0.40}As/GaAs$ self-assembled quantum dots*, Phys. Rev. B **65** (2002), 041308(R)(1–4).

[BGL88] Dieter Bimberg, Marius Grundmann, and Nikolai Ledentsov, *Quantum dot heterostructures*, Wiley, Chichester, UK, 1988.

[BGT+99] J. Brendel, N. Gisin, W. Tittel, and H. Zbinden, *Pulsed energy-time entangled twin-photon source for quantum communication*, Phys. Rev. Lett. **82** (1999), 2594–2597.

[BHH+01] M. Bayer, P. Hawrylak, K. Hinzer, S. Fafard, M. Korkusinski, Z. R. Wasilewski, O. Stern, and A. Forchel, *Coupling and entangling of quantum states in quantum dot molecules*, Science **291** (2001), 451–453.

[BKB+02] A. Beveratos, S. Kühn, R. Brouri, T. Gacoin, J.-P. Poizat, and P. Grangier, *Room temperature stable single-photon source*, Eur. Phys. J. D **18** (2002), 191–196.

[BKM+01] L. Besombes, K. Kheng, L. Masal, and H. Mariette, *Acoustic phonon broading mechanism in single quantum dot emission*, Phys. Rev. B **63** (2001), 155307(1–5).

[BLH01] D. Birkedal, K. Leosson, and J. M. Hvam, *Long lived coherence in self-assembled quantum dots*, Phys. Rev. Lett. **87** (2001), 227401(1–4).

[BLS+01] P. Borri, W. Langbein, S. Schneider, U. Woggon, R. L. Sellin, D. Ouyang, and D. Bimberg, *Ultralong dephasing time in InGaAs quantum dots*, Phys. Rev. Lett. **87** (2001), 157401(1–4).

[BLT+99] Christian Brunel, Brahim Lounis, Philippe Tamarat, and Michel Orrit, *Triggered source of single photons based on controlled single molecule fluorescence*, Phys. Rev. Lett. **83** (1999), 2722–2725.

[BOS+02] M. Bayer, G. Ortner, O. Stern, A. Kuther, A. A. Gorbunov, A. Forchel, P. Hawrylak, S. Fafard, K. Hinzer, T. L. Reinecke, S. N. Walck, J. P. Reithmaier, F. Klopf, and F. Schäfer, *Fine structure of neutral and charged* excitons in self-assembled In(Ga)As/(Al)GaAs quantum dots, Phys. Rev. B **65** (2002), 195315(1–23).

[BPK⁺04] M. H. Baier, E. Pelucchi, E. Kapon, S. Varoutsis, M. Gallart, I. Robert-Philip, and I. Abram, *Single photon emission from site-controlled pyramidal quantum dots*, Appl. Phys. Lett. **84** (2004), 648–650.

[BPM⁺97] Dik Bouwmeester, Jian-Wei Pan, Klaus Mattle, Manfred Eibl, Harald Weinfurter, and Aanton Zeilinger, *Experimental quantum teleportation*, Nature **390** (1997), 575–579.

[BSP⁺00] Oliver Benson, Charles Santori, Matthew Pelton, and Yoshihisa Yamamoto, *Regulated and entangled photons from a single quantum dot*, Phys. Rev. Lett. **84** (2000), 2513–2516.

[BW99] Max Born and Emil Wolf, *Principles of optics*, Cambridge University Press, 1999.

[BWH⁺00] P. G. Blome, M. Wenderoth, M. Hübner, R. G. Ulbrich, J. Porsche, and F. Scholz, *Temperature-dependent linewidth of single InP/Ga$_x$In$_{1-x}$P quantum dots: Interaction with surrounding charge configurations*, Phys. Rev. B **61** (2000), 8382–8387.

[BWS⁺99] G. Bacher, R. Weigand, J. Seufert, V. D. Kulakovskii, N. A. Gippius, A. Forchel, K. Leonardi, and D. Hommel, *Biexciton versus exciton lifetime in a single semiconductor quantum dot*, Phys. Rev. Lett. **83** (1999), 4417–4420.

[BZK⁺03] C. Braig, P. Zarda, C. Kurtsiefer, and H. Weinfurter, *Experimental demonstartion of complementarity with single photons*, Appl. Phys. B **76** (2003), 113–116.

[CSP⁺94] N. Carlsson, W. Seifert, A. Petersson, P. Castrillo, M.-E. Pistol, and L. Samuelson, *Study of the two-dimensional—three-dimensional growth mode transition in metalorganic vapor phase epitaxy of GaInP/InP quantum-sized structures*, Appl. Phys. Lett. **65** (1994), 3093–3095.

[CVZ⁺98] Isaak L. Chuang, Lieven M. K. Vandersypen, Xinlan Zhou, Debbie W. Leung, and Seth Lloyd, *Experimental realization of a quantum algorithm*, Nature **393** (1998), 143–146.

[CWL⁺04] D. Chithrani, R. L. Williams, J. Lefebvre, P. J. Poole, and G. C. Aers, *Optical spectroscopy of single, site-selected, InAs/InP self-assembled quantum dots*, Appl. Phys. Lett. **84** (2004), 978–980.

[DGE+00] E. Dekel, D. Gershoni, E. Ehrenfreund, J. M. Garcia, and P. M. Petroff, *Carrier-carrier correlations in an optically excited single semiconductor quantum dot*, Phys. Rev. B **61** (2000), 11009–11020.

[DiV00] David P. DiVincenzo, *The physical implementation of quantum computation*, Fortschr. Phys. **48** (2000), 771–783.

[Eke91] Artur K. Ekert, *Quantum cryptography based on Bell's theorem*, Phys. Rev. Lett. **67** (91), 661–663.

[EOT80] A. I. Ekimov, A. A. Onushchenko, and V. A. Tsekhomskii, *Exciton light absorption by CuCl microcrystals in glass matrix*, Sov. Glass Phys. Chem. **6** (1980), 511–512.

[FBA+04] T. Flissikowski, A. Betke, I. A. Akimov, and F. Henneberger, *Two-photon coherent control of a single quantum dot*, Phys. Rev. Lett. **92** (2004), 227401(1–4).

[Fey82] Richard P. Feynman, *Simulating physics with computers*, Int. J. Theor. Phys. **21** (1982), 467.

[FHL+01] T. Flissikowski, A. Hundt, M. Lowisch, M. Rabe, and F. Henneberger, *Photon beats from a single semiconductor quantum dot*, Phys. Rev. Lett. **86** (2001), 3172–3175.

[Fli04] Timur Flissikowski, *Coherence properties of single self-assembled quantum dots*, Ph.D. thesis, Humboldt-Universität zu Berlin, 2004.

[FYL00] M. Fleischhauer, S. F. Yelin, and M. D. Lukin, *How to trap photons? Storing single-photon quantum states in collective atomic excitations*, Opt. Comm. **179** (2000), 394–410.

[GBS01] Stephan Götzinger, Oliver Benson, and Vahid Sandoghdar, *Towards controlled coupling between a high-Q whispering-gallery mode and a single nanoparticle*, Appl. Phys. B **73** (2001), 825–828.

[Gla63] Roy J. Glauber, *Quantum theory of optical coherence*, Physical Review **130** (1963), 2529–2539.

[Göt03] Stephan Götzinger, *Controlled coupling of a single nanoparticle to a high-Q microsphere resonator*, Ph.D. thesis, Humboldt-Universität zu Berlin, 2003.

[GRA86] P. Grangier, G. Roger, and A. Aspect, *Experimental evidence for a photon anticorrelation effect on a beamsplitter*, Europhys. Lett. **1** (1986), 173–179.

[GRL+03]   Stephan Gulde, Mark Riebe, Gavin P. T. Lancaster, Christoph Becher,
           Jürgen Eschner, Hartmut Häffner, Ferdinand Schmidt-Kaler, Isaac L.
           Chuang, and Rainer Blatt, *Implementation of the Deutsch-Jozsa algorithm
           on an ion-trap quantum computer*, Nature **421** (2003), 48–50.

[Gro97]   Lev K. Grover, *Quantum mechanics helps in searching for a needle in a
           haystack*, Phys. Rev. Lett. **79** (1997), 325–328.

[GRT+02]   Nicolas Gisin, Grégoire Ribordy, Wolfgang Tittel, and Hugo Zbinden, *Quan-
           tum cryptography*, Rev. Mod. Phys. **74** (2002), 145–195.

[GSG+98]   J. M. Gérard, B. Sermage, B. Gayral, B. Legrand, E. Costard, and
           V. Thierry-Mieg, *Enhanced spontaneous emission by quantum boxes in a
           monolithic optical microcavity*, Phys. Rev. Lett. **81** (1998), 1110–1113.

[HBE+97]   J. T. Höffges, H. W. Baldauf, T. Eichler, S. R. Helmfrid, and H. Walther,
           Heterodyne measurement of the fluorescent radiation of a single trapped ion,
           Opt. Comm. **133** (1997), 170–174.

[Hec02]   Jeff Hecht, *Understanding fibre optics*, Prentice-Hall, London, 2002.

[HOM87]   C. K. Hong, Z. Y. Ou, and L. Mandel, *Measurement of subpicosecond time
           intervals between two photons by interference*, Phys. Rev. Lett. **59** (1987),
           2044–2046.

[HT56]   R. Hanbury Brown and R. Q. Twiss, *A test of a new type of stellar inter-
           ferometer on Sirius*, Nature **178** (1956), 1046–1048.

[HVG+03]   J. Hours, S. Varoutsis, M. Gallart, J. Bloch, I. Robert-Philip, A. Cavanna,
           I. Abram, F. Laruelle, and J. M. Gérard, *Single photon emission from in-
           dividual GaAs quantum dots*, 2003, pp. 2206–2208.

[JHH+03]   M. Johansson, U. Håkanson, M. Holm, J. Persson, T. Sass, J. Johansson,
           C. Pryor, L. Montelius, W. Seifert, L. Samuelson, and M.-E. Pistol, *Corre-
           lation between overgrowth morphology and optical properties of single self-
           assembled InP quantum dots*, Phys. Rev. B **68** (2003), 125303(1–9).

[JN58]   L. Jánossy and Zs. Náray, *Investigation into interference phenomena at
           extremely low light intensities by means of a large Michelson interferometer*,
           Nuovo Cimento **9** (1958), 588–598.

[JPSG+]   M. Jetter, V. Pérez-Solórzano, A. Gröning, M. Ubl, H. Gräbeldinger, and
           H. Schweizer, *Selective growth of GaInN quantum dot structures*, J. Cryst.
           Growth, to be published.

[JSW+00]   Thomas Jennewein, Christoph Simon, Gregor Weihs, Harald Weinfurter, and Anton Zeilinger, *Quantum cryptography with entangled photons*, Phys. Rev. Lett. **84** (2000), 4729–4732.

[JVP+03]   F. Jelezko, A. Volkmer, I. Popa, K. K. Rebane, and J. Wrachtrup, *Coherence length of photons from a single quantum system*, Phys. Rev. A **67** (2003), 041802(R)(1–4).

[KCV+02]   C. Kammerer, G. Cassabois, C. Voisin, M. Perrin, C. Delalande, Ph. Roussignol, and J. M. Gérard, *Interferometric correlation spectroscopy in single quantum dots*, Appl. Phys. Lett. **81** (2002), 2737–2739.

[KDM77]   H. J. Kimble, M. Dagenais, and L. Mandel, *Photon antibunching in resonance fluorescence*, Phys. Rev. Lett. **39** (1977), 691–695.

[KHR02]   A. Kuhn, M. Hennrich, and G. Rempe, *Deterministic single-photon source for distributed quantum networking*, Phys. Rev. Lett. **89** (2002), 067901(1–4).

[KLH+04]   Matthias Keller, Birgit Lange, Kazuhiro Hayasaka, Wolfgang Lange, and Herbert Walther, *Continuous generation of single photons with controlled waveform in an ion-trap cavity system*, Nature **431** (2004), 1075–1078.

[KLM01]   E. Knill, R. Laflamme, and G. J. Milburn, *A scheme for efficient quantum computation with linear optics*, Nature **409** (2001), 46–52.

[KMW+95]   Paul G. Kwiat, Klaus Mattle, Harald Weinfurter, Anton Zeilinger, Alexander V. Sergienko, and Yanhua Shih, *New high-intensity source of polarization-entangled photon pairs*, Phys. Rev. Lett. **75** (1995), 4337–4341.

[KMZ+00]   C. Kurtsiefer, S. Mayer, P. Zarda, and H. Weinfurter, *Stable solid-state source of single photons*, Phys. Rev. Lett. **85** (2000), 290–293.

[KVC+02]   C. Kammerer, C. Voisin, G. Cassabois, C. Delalande, Ph. Roussignol, F. Klopf, J. P. Reithmaier, A. Forchel, and J. M. Gérard, *Line narrowing in single semiconductor quantum dots: Toward the control of environment effects*, Phys. Rev. B **66** (2002), 041306(R)(1–4).

[KZH+02]   C. Kurtsiefer, P. Zarda, M. Halder, H. Weinfurter, P. M. Gorman, P. R. Tapster, and J. G. Rarity, *A step towards global key distribution*, Nature **419** (2002), 450.

[KZM+01]   Christian Kurtsiefer, Patrick Zarda, Sonja Mayer, and Harald Weinfurter, *The breakdown flash of silicon avalanche photodiodes — backdoor for eavesdropper attacks?*, J. Mod. Phys **48** (2001), 2093–2047.

[Lam95]   W. E. Lamb, *Anti-photon*, Appl. Phys. B **60** (1995), 77–84.

[Las]   Laser Components GmbH, Germany, *Single photon counting module – SPCM-AQR series specifications.*

[LHA⁺01]   Alexander I. Lvovsky, Hauke Hansen, Thomas Aichele, Oliver Benson, Jürgen Mlynek, and Stephan Schiller, *Quantum state reconstruction of the single-photon Fock state*, Phys. Rev. Lett. **87** (2001), 050402(1–4).

[LM00]   B. Lounis and W. E. Moerner, *Single photons on demand from a single molecule at room temperature*, Nature **407** (2000), 491–493.

[LRG⁺02]   D. Litvinov, A. Rosenauer, D. Gerthsen, P. Kratzert, M. Rabe, and F. Henneberger, *Influence of the growth procedure on the Cd distribution in CdSe/ZnSe heterostructures: Stranski-Krastanov versus two-dimensional islands*, Appl. Phys. Lett. **81** (2002), 640–642.

[Lüt00]   Norbert Lütkenhaus, *Security against individual attacks for realistic quantum key distribution*, Phys. Rev. A **61** (2000), 052304(1–10).

[LWS⁺03]   X. Li, Y. Wu, D. Steel, D. Gammon, T. H. Stievater, D. S. Katzer, D. Park, C. Piermarocchi, and L. J. Sham, *An all-optical quantum gate in a semiconductor quantum dot*, Science **301** (2003), 809–811.

[Mas02]   Yasuaki Masumoto, *Coherent spectroscopy of semiconductor quantum dots*, J. Luminescence **100** (2002), 191–208.

[MBB⁺04]   J. McKeever, A. Boca, A. D. Boozer, R. Miller, J. R. Buck, A. Kuzmich, and H. J. Kimble, *Deterministic generation of single photons from one atom trapped in a cavity*, Science **303** (2004), 1992–1994.

[MHM⁺00]   Jens Michaelis, Christian Hettich, Jürgen Mlynek, and Vahid Sandoghdar, *Optical microscopy using a single-molecule light source*, Nature **405** (2000), 325–328.

[Mic81]   Albert Michelson, *The relative motion of the earth and the luminiferous aether*, Am. J. Sci **22** (1881), 120–129.

[MIM⁺00]   P. Michler, A. Imamoğlu, M. D. Mason, P. J. Carson, G. F. Strouse, and S. K. Buratto, *Quantum correlation among photons from a single quantum dot at room temperature*, Nature **406** (2000), 968–970.

[Mir04]   Richard P. Mirin, *Photon antibunching at high temperature from a single InGaAs/GaAs quantum dot*, Appl. Phys. Lett. **84** (2004), 1260–1262.

[MM87]   Albert Michelson and Edward Morley, *On the relative motion of the earth and the luminiferous ether*, Am. J. Sci **34** (1887), 333–345.

[MNB93]   C. B. Murray, D. J. Noms, and M. G. Bawendi, *Synthesis and characterization of nearly monodisperse CdE (E = S, Se, Te) semiconductor nanocrystallites*, J. Am. Chem. Soc. **115** (1993), 8706–8715.

[MRG⁺01]   E. Moreau, I. Robert, J. M. Gérard, I. Abram, L. Manin, and V. Thierry-Mieg, *Single-mode solid-state single photon source based on isolated quantum dots in pillar microcavities*, Appl. Phys. Lett. **79** (2001), 2865–2867.

[MRM⁺01]   E. Moreau, I. Robert, L. Manin, V. Thierry-Mieg, J. M. Gérard, and I. Abram, *Quantum cascade of photons in semiconductor quantum dots*, Phys. Rev. Lett. **87** (2001), 183601(1–4).

[MT02]   Yasuaki Masumoto and Toshihide Takagahara, *Semiconductor quantum dots. Physics, spectroscopy and applications*, Springer-Verlag, 2002.

[MW95]   Leonard Mandel and Emil Wolf, *Optical coherence and quantum optics*, Cambridge University Press, 1995.

[MZ04]   E. A. Muljarov and R. Zimmermann, *Dephasing in quantum dots: Quadratic coupling to acoustic phonons*, Phys. Rev. Lett. **93** (2004), 237401(1–4).

[NVL97]   Aleksandr Noy, Dmitri V. Vezenov, and Charles M. Lieber, *Chemical force microscopy*, Ann. Rev. Mat. Sci. **27** (1997), 381–421.

[OPH⁺93]   D. Oberhauser, K.-H. Pantke, J. M. Hvam, G. Weimann, and C. Klingshirn, *Exciton scattering in quantum wells at low temperatures*, Phys. Rev. B **47** (1993), 6827–6830.

[PHP⁺03]   J. Persson, M. Holm, C. Pryor, D. Hessmann, W. Seifert, L. Samuelson, and M.-E. Pistol, *Optical and theoretical investigations of small InP quantum dots in $Ga_xIn_{1-x}P$*, Phys. Rev. B **67** (2003), 035320(1–7).

[PSV⁺02]   M. Pelton, C. Santori, J. Vučković, B. Zhang, G. S. Solomon, J. Plant, and Y. Yamamoto, *Efficient source of single photons: A single quantum dot in a micropost microcavity*, Phys. Rev. Lett. **89** (2002), 233602(1–4).

[Pur46]   E. M. Purcell, *Spontaneous emission probabilities at radio frequencies*, Phys. Rev. **69** (1946), 681.

[RLH98]   M. Rabe, M. Lowisch, and F. Henneberger, *Self-assembled CdSe quantum dots formation by thermally activated surface reorganization*, J. Cryst. Growth **184/185** (1998), 248–253.

[RMG⁺01] D. V. Regelman, U. Mizrahi, D. Gershoni, E. Ehrenfreund, W. V. Schoen-
feld, and P. M. Petroff, *Semiconductor quantum dot: A quantum light source
of multicolor photons with tunable statistics*, Phys. Rev. Lett. **87** (2001),
257401(1–4).

[RSL⁺04] J. P. Reithmaier, G. Sęk, A. Löffler, C. Hofmann, S. Kuhn, S. Reitzenstein,
L. V. Keldysh, V. D. Kulakovskii, T. L. Reinecke, and A. Forchel, *Strong
coupling in a single quantum dot–semiconductor microcavity system*, Nature
432 (2004), 197–200.

[RSS69] Geo T. Reynolds, K. Spartalian, and D. B. Scarl, *Interference effects produced by single photons*, Nuovo Cimento **61 B** (1969), 355–364.

[SFP⁺02] Charles Santori, David Fattal, Matthew Pelton, Glenn S. Solomon, and
Yoshihisa Yamamoto, *Polarization-correlated photon pairs from a single
quantum dot*, Phys. Rev. B **66** (2002), 045308(1–4).

[SFV⁺02] Charles Santori, David Fattal, Jelena Vučković, Glenn S. Solomon, and
Yoshihisa Yamamoto, *Indistinguishable photons from a single-photon device*,
Nature **419** (2002), 594–597.

[SFV⁺04] Charles Santori, David Fattal, Jelena Vučković, Glenn S. Solomon, Edo
Waks, and Yoshihisa Yamamoto, *Submicrosecond correlations in photoluminescence from InAs quantum dots*, Phys. Rev. B **69** (2004), 205324(1–8).

[Sho97] P. W. Shor, *Polynomial-time algorithms for prime factorization and discrete
logarithms on a quantum computer*, Proc. 35th Ann. Symp. on Found. of
Comp. Science, IEEE Press, 1997, pp. 124–134, see also quant–ph/9508027.

[SMP⁺02] K. Sebald, P. Michler, T. Passow, D. Hommel, G. Bacher, and A. Forchel,
Single-photon emission of CdSe quantum dots at temperatures up to 200 K,
Appl. Phys. Lett. **81** (2002), 2920–2922.

[SP04] Christoph Simon and Jean-Philippe Poizat, *Creating single time-bin entangled photon pairs*, arXiv:quant-ph (2004), 0409100(1–4).

[SPS⁺01] Charles Santori, Matthew Pelton, Glenn S. Solomon, Yseulte Dale, and
Yoshihisa Yamamoto, *Triggered single photons from a quantum dot*, Phys.
Rev. Lett. **86** (2001), 1502–1505.

[SPY01] Glenn Solomon, Matthew Pelton, and Yoshihisa Yamamoto, *Single-mode
spontaneous emission from a single quantum dot in a three-dimensional microcavity*, Phys. Rev. Lett. **86** (2001), 3903–3906.

[ST91] Bahaa E. A. Saleh and Malvin Carl Teich, *Fundamentals of photonics*, Wiley, New York, 1991.

[Tay09]   G. I. Taylor, Proc. Cam. Phil. Soc. Math. Phys. Sci. **15** (1909), 114.

[TSN⁺00]   Y. Toda, T. Sugimoto, M. Nishioka, and Y. Arakawa, *Near-field coherent excitation spectroscopy of InGaAs/GaAs self-assembled quantum dots*, Appl. Phys. Lett. **76** (2000), 3887–3889.

[UBM⁺04]   S. M. Ulrich, M. Benyoucef, P. Michler, N. Baer, P. Gartner, F. Jahnke, M. Schwab, H. Kurtze, M. Bayer, S. Fafard, and Z. Wasilewski, *Correlated photon-pair emission from a charged single quantum dot*, arXiv:cond-mat/0407441v1 (2004), 1–5.

[UOO⁺93]   M. Ukita, H. Okuyama, M. Ozawa, A. Ishibashi, K. Akimoto, and Y. Mori, *Refractive indices of ZnMgSSe alloys lattice matched to GaAs*, Appl. Phys. Lett. **63** (1993), 2082–2084.

[USM⁺03]   S. M. Ulrich, S. Strauf, P. Michler, G. Bacher, and A. Forchel, *Triggered polarization-correlated photon pairs from a single CdSe quantum dot*, Appl. Phys. Lett. **83** (2003), 1848–1850.

[VMR⁺96]   M. Vollmer, E. J. Mayer, W. W. Rühle, A. Kurtenbach, and K. Eberl, *Exciton relaxation dynamics in quantum dots with strong confinement*, Phys. Rev. B **54** (1996), R17292–R17295.

[WIS⁺02]   Edo Waks, Kyo Inoue, Charles Santori, David Fattal, Jelena Vučković, Glenn S. Solomon, and Yoshihisa Yamamoto, *Quantum cryptography with a photon turnstile*, Nature **420** (2002), 762.

[WZ82]   W. K. Wooters and W. H. Zurek, *A single quantum cannot be cloned*, Nature **299** (1982), 802–803.

[XWC04]   X. Xu, D. A. Williams, and J. R. A. Cleaver, *Electrically pumped single-photon sources in lateral p-i-n junctions*, Appl. Phys. Lett. **85** (2004), 3238–3240.

[Yao85]   Takafumi Yao, *Characterization of ZnSe grown by molecular-beam epitaxy*, J. Cryst. growth **72** (1985), 31–40.

[YHC⁺02]   Yang Yu, Siyuan Han, Xi Chu, Shih-I Chu, and Zhen Wang, *Coherent temporal oscillations of macroscopic quantum states in a Josephson junction*, Science **296** (2002), 889–892.

[YKS⁺02]   Z. Yuan, B. E. Kardynal, R. M. Stevenson, A. J. Shields, C. J. Lobo, K. Cooper, N. S. Beattie, D. A. Ritchie, and M. Pepper, *Electrically driven single-photon source*, Science **295** (2002), 102–105.

[YPA⁺03]   T. Yamamoto, Yu. A. Pashkin, O. Astafiev, Y. Nakamura, and J. S. Tsai, *Demonstration of conditional gate operation using superconducting charge qubits*, Nature **425** (2003), 941–944.

[YSH⁺04]   T. Yoshie, A. Scherer, J. Hendrickson, G. Khitrova, H. M. Gibbs, G. Rupper, C. Ell, O. B. Shchekin, and D. G. Deppe, *Vacuum Rabi splitting with a single quantum dot in a photonic crystal nanocavity*, Nature **432** (2004), 200–203.

[ZAB04]   Valéry Zwiller, Thomas Aichele, and Oliver Benson, *Single-photon Fourier spectroscopy of excitons and biexcitons in single quantum dots*, Phys. Rev. B **69** (2004), 165307(1–4).

[ZAH⁺04]   Valéry Zwiller, Thomas Aichele, Fariba Hatami, W. Ted Masselink, and Oliver Benson, *Growth of single quantum dots on preprocessed structures: single photon emitters on a tip*, Appl. Phys. Lett. **86** (2004), 091911(1–3).

[ZAP⁺03]   Valéry Zwiller, Thomas Aichele, Jonas Persson, Lars Samuelson, and Oliver Benson, *Generating visible single photons on demand with single InP quantum dots*, Appl. Phys. Lett. **82** (2003), 1509–1511.

[ZB02]   Valéry Zwiller and Gunnar Björk, *Improved light extraction from emitters in high refractive index materials using solid immersion lenses*, J. Appl. Phys. **92** (2002), 660–665.

[ZBJ⁺01]   Valéry Zwiller, Hans Blom, Per Jonsson, Nikolay Panev, Sören Jeppesen, Tedros Tsegaye, Edgard Goobar, Mats-Erik Pistol, Lars Samuelson, and Gunnar Björk, *Single quantum dots emit single photons at a time: Antibunching experiments*, Appl. Phys. Lett. **78** (2001), 2476–2478.

[ZPH⁺99]   Valéry Zwiller, Mats-Erik Pistol, Dan Hessmann, Rolf Cederström, Werner Seifert, and Lars Samuelson, *Time-resolved studies of single semiconductor quantum dots*, Phys. Rev. B **59** (1999), 5021–5025.

[ZVB04]   Alessandro Zavatta, Silvia Viciani, and Marco Bellini, *Quantum-to-classical transition with single-photonadded coherent states of light*, Science **306** (2004), 660–662.

[ZZB⁺04]   V. Zwiller, C. Zinoni, G. Buchs, B. Alloing, and A. Fiore, to be submitted, 2004.

Acknowledgements

First of all I thank Prof. Oliver Benson for opening my eye to the world of nano-optics and for giving me the opportunity to work on this interesting experiment. Oliver always had time for consulting him about every problem and still allowed me much autonomy for my work.

I am also grateful to Prof. Jürgen Mlynek for his valuable support since I started in his group in Konstanz. I appreciated very much having worked in his exceptional group.

I am deeply indebted to Val Zwiller for a very fruitful team work even beyond his time in Berlin. I learned a lot from him about quantum dots and nano structures. He inspired this work with many ideas by countless interesting discussions.

I enjoyed very much to work with Matthias Scholz during the last year, and I wish him all the best in his experiment. I thank Gaël Reinaudi for his indispensable work on the quantum cryptography setup and for bringing a cheerful French atmosphere into the Single-Photon Lab. I am grateful to Jonas Persson for the cooperation during his visits in Berlin.

This work would not have been possible without the high-quality quantum dot samples I have explored. I thank Prof. Werner Seifert from Lund University for providing the InP sample with which most of the experiments in this thesis were performed. The CdSe quantum dot sample was gratefully provided by Ilya Akimov and Prof. Fritz Henneberger. I thank Fariba Hatami and Prof. W. Ted Masslink for the growth of InP quantum dots on the tips. The tip structures were processed with grateful assistance of E. Wiebicke from Paul Drude Institute.

I appreciate many inspiring discussions with Ulrike Herzog, Prof. Roland Zimmermann and Ilya Akimov.

From the very beginning I spent a lot of joyful time with Stephan Götzinger. I thank him for not killing me when eating Butter unner dr Lewwerworscht. I liked very much the time I spent with Leonardo Menezes (maybe apart from that *one* day after). Holger Müller always found time for discussing and solving problems or for just having a date for a beer. Andrea Mazzei, Evgeny Kovalchuk and Stefan Schietinger enriched very much my life in- and outside the lab. For the cheerful atmosphere at Hausvogteiplatz I am grateful to the Nanos Kirstin Wohlfart, Felix Müller, Ounsi El-Daif, Franz Boczianowski,

Sven Ramelow, Gesine Steudle, Hanna Krauter, Björn Lauritzen and to the Quantum Metrologists and Modern Opticians.

I also want to thank all my colleagues in Konstanz. Particularly Prof. Vahid Sandoghdar and his group members Christian Hettich, Patrick Kramper, Hannes Schniepp, Thomas Kalkbrenner, Ilja Gerhard, Jan Zitzmann and Lavinia Rogobete helped me a lot with my first steps in the field of nano-optics.

Klaus Palis was a great help in building up the Berlin lab and a good advisor in technical questions. I also thank Stefan Eggert and Stefan Hahn in Konstanz for engineering assistance.

Matthias, Katrin and Paddy spent a lot of time in reading and correcting this thesis and eliminated many misttakés.

I am thankful to my friends in Berlin and *am See* who also contributed to the nice time I had during my graduation. I enjoy very much visionary discussions with Paddy Polzer.

My warmest thank-you is dedicated to my parents Christa and Hans-Karl and to my sister Katrin who supported me all the time in a variety of ways.

Publications

Most of this work is already published in the following articles:

- Thomas Aichele, Ulrike Herzog, Matthias Scholz and Oliver Benson,
  Single photon generation and simultaneous observation of wave and particle properties,
  Proc. AIP on Foundations of Probability and Physics **750**, 35 (2004),
  see also arXiv:quant-ph/0410112v2.

- Thomas Aichele, Gaël Reinaudi and Oliver Benson,
  Separating Cascaded Photons From a Single Quantum Dot: Demonstration of Multiplexed Quantum Cryptography,
  Phys. Rev. B **70**, 235329 (2004).

- Valéry Zwiller, Thomas Aichele, Fariba Hatami, W. Ted Masselink and Oliver Benson,
  Growth of single quantum dots on preprocessed structures: Single photon emitters on a tip,
  Appl. Phys. Lett. **86**, 091911 (2005).

- Jonas Persson, Thomas Aichele, Valéry Zwiller, Lars Samuelson and Oliver Benson,
  Three photon cascade from a single semiconductor quantum dot,
  Phys. Rev. B **69**, 233314, 1–4 (2004).

- Valéry Zwiller, Thomas Aichele and Oliver Benson,
  Single Photon Fourier Spectroscopy of Excitons and Biexcitons in Single Quantum Dots,
  Phys. Rev. B **69**, 165307, 1–4 (2004).

- Thomas Aichele, Valéry Zwiller and Oliver Benson,
  Visible single photon generation from semiconductor quantum dots,
  New J. Phys. **6**, 90, 1–13 (2004).

- Valéry Zwiller, Thomas Aichele and Oliver Benson,
  Quantum optics with single quantum dot devices,
  New J. Phys. **6**, 96, 1–23 (2004).

- Thomas Aichele, Valéry Zwiller, Oliver Benson, Ilya Akimov and Fritz Henneberger,
  Single CdSe quantum dots for high bandwidth single photon generation,
  J. Opt. Soc. Am. B **20**, 2189–2192 (2003).

- Valéry Zwiller, Thomas Aichele, Werner Seifert, Jonas Persson and Oliver Benson,
  Generating visible single photons on demand with single InP quantum dots,
  Appl. Phys. Lett. **82**, 1509–1511 (2003).

Further publications:

- Alex I. Lvovsky and Thomas Aichele,
  Conditionally prepared photon and quantum imaging,
  Proc. SPIE **5551**, 1–6 (2004)

- Thomas Aichele, Alex I. Lvovsky and Stephan Schiller,
  Optical mode characterization of single photons prepared by means of conditional measurements on a biphoton state,
  Eur. Phys. J. D **18**, 237–245 (2002).

- Alex I. Lvovsky, Hauke Hansen, Thomas Aichele, Oliver Benson, Jürgen Mlynek and Stephan Schiller,
  Quantum State Reconstruction of the Single-Photon Fock State,
  Phys. Rev. Lett. **87**, 050402, 1–4 (2001),
  see also: R. Slayton, *Golfing with a Single Photon,* Phys. Rev. Focus to PRL **87**, 050402 (2001).

- Hauke Hansen, Thomas Aichele, Christian Hettich, Peter Lodahl, Alex I. Lvovsky, Jürgen Mlynek and Stephan Schiller,
  Ultrasensitive pulsed, balanced homodyne detector: application to time-domain quantum measurements
  Opt. Lett. **26**, 1714–1716 (2001).

Selbständigkeitserklärung

Hiermit erkläre ich, die vorliegende Arbeit selbständig ohne fremde Hilfe verfasst und nur die angegebene Literatur und Hilfsmittel verwendet zu haben.

Ich habe mich anderwärts nicht um einen Doktorgrad beworben und besitze einen entsprechenden Doktorgrad nicht.

Ich erkläre die Kenntnisnahme der dem Verfahren zugrunde liegenden Promotionsordnung der Mathematisch-Naturwissenschaftlichen Fakultät I der Humboldt-Universität zu Berlin.

Berlin, den 7. Dezember 2004

Thomas Aichele